STEVEN LEVY

INSANELY GREAT

THE LIFE AND TIMES OF MACINTOSH, THE COMPUTER THAT CHANGED EVERYTHING

PENGUIN BOOKS

To levyd @ uncwil.edu

PENGUIN BOOKS

Published by the Penguin Group
Penguin Books Ltd, 27 Wrights Lane, London W8 5TZ, England
Penguin Books USA Inc., 375 Hudson Street, New York, New York 10014, USA
Penguin Books Australia Ltd, Ringwood, Victoria, Australia
Penguin Books Canada Ltd, 10 Alcorn Avenue, Toronto, Ontario, Canada M4V 3B2
Penguin Books (NZ) Ltd, 182–190 Wairau Road, Auckland 10, New Zealand

Penguin Books Ltd, Registered Offices: Harmondsworth, Middlesex, England

First published by Viking 1994
Published with minor revisions and an Afterword in Penguin Books 1995
1 3 5 7 9 10 8 6 4 2

Grateful acknowledgment is made for permission to use the voiceover from the
'1984' television commercial introducing the Macintosh computer.
By permission of Apple Computer, Inc.

The afterword is adapted from Mr Levy's two column series on the PowerMac,
which appeared in the May 1994 and June 1994 issues of *Macworld*.

Printed in England by Clays Ltd, St Ives plc

ACKNOWLEDGMENTS

I owe a huge debt to my sources at Apple, the third-party community, and the wider Macintosh world at large. For ten years they have patiently answered my queries, explained technical issues to me, and more often than not provided me with enlightening conversation that illuminated my thinking about matters both Mac and non-Mac. I am especially grateful to the help far beyond the call of duty to Mac Team members Bill Atkinson, Steve Capps, Andy Hertzfeld, Joanna Hoffman, and Susan Kare, who have always been there for me when I needed them. They also helped produce a fairly impressive computer, which of course was the tool I used to generate this manuscript. It was also the instrument used by Viking to design this book, lay it out, and send it straight to the printer.

Thanks also to the normally thankless public relations people who have linked me to these sources throughout the years, often without complaining about the result. These include the beleaguered PR staffs at Apple, Aldus, Regis McKenna, Microsoft, Waggoner-Edstrom, and dozens of third-party vendors and agencies. Particular thanks to Jane Anderson, Andy Cunningham, Marty Taucher, Kate Paisley, Barbara Krause, and Yolanda Davis.

My excursions in computer journalism have been greatly aided by incisive and supportive editors: Rich Friedman of the late *Popular Computing*; David Rosenthal and, later, Bob Love of *Rolling Stone*; and at *Macworld*, three aces in a row—Dan Farber, Nancy Dunn, and Deborah Branscum. (Deborah was also my editor for the two-column series on the PowerMac that appeared in *Macworld* in the May and June editions of 1994. Those columns have been adapted into the Afterword which appears at the end of this paperback edition.) I've also benefited from expertise and advice from many other *Macworld* colleagues, beginning with editors-in-chief David Bunnell, Jerry Borrell, and Adrian Mello, and down through the masthead—sadly, too many names to list here.

The manuscript benefited from comments by Deborah Branscum, Teresa Carpenter, Andy Hertzfeld, Joanna Hoffman, Susan Kare, John Markoff, and Larry Tesler. For this paperback edition, I'd like to thank those readers who took it upon themselves to notify me of some random bugs in the original: Ted Bear, Howard J. Frolich, Mike Morton, Ernst J. Oud, Jef Raskin, Peter Stoller, and David Wells. (If I missed someone, I apologize.) The usual disclaimer applies, though: All blunders are mine.

Thanks, too, to my agent Flip Brophy for constant support; and to editor Pam Dorman, whose enthusiasm for the project lured me to Viking. And, as always, my gratitude and love to Teresa and Andrew Max.

The world has arrived at an age of cheap complex devices of great reliability; and something is bound to come of it.

—VANNEVAR BUSH

1

What I first remember was the light.

It was November 1983, and after some Byzantine negotiations, I was admitted to the inner sanctum—a low-slung building in Cupertino, California, containing the most whispered-about secret since the Enigma, or at least since Who shot J.R.? Upon confirming my identity, the receptionist directed me to a small conference room named after a French painter. A short, energetic woman in a suede jumpsuit entered carrying an awkward canvas bag. She set the bag on the table, unzipped the top and reached in, grabbing something by a recessed handle.

The shape is now a familiar component of our culture, as instantly recognizable as a Volkswagen or a Coke bottle. Back then, I had never seen anything like it. All I knew was its name: Macintosh. And that it was supposed to change the world. It certainly looked different.

In about thirty seconds, the woman had everything plugged in and connected. She reached behind it and turned it on. The disk drive ground and whirred. And the small screen turned milky white. In the middle was a sharp little machine self-portrait, with a blinking question mark inside on the screen inside the screen. Then the disk drive whirred once more and the question mark evaporated. In its place was a happy face. Macintosh was happy.

I was witnessing a revolution.

Until that moment, when one said a computer screen "lit up," some literary license was required. Unless the display was something from a graphics program or a game, the background on a monitor was invariably black, providing a contrast to the phosphorescent green (sometimes white) letters. Reading text off a computer screen had the feel of staring into the flat bottom part of those toy fortune-telling Eight Balls, where you'd ask the thing a question, turn it upside down, and a cryptic answer would dreamily drift into view. Everyone who used computers considered this one of the standard discomforts: it did hurt your eyes if you stared too long. But we were so accustomed to it that we hardly even thought to conceive otherwise. We simply hadn't seen the light.

I saw it that day. I also saw many things I didn't know a computer could do. By the end of the demonstration, I began to understand that these were things a computer *should* do. There was a better way.

On that day in November, I met the people who created that machine. They were groggy and almost giddy from three years of creation. Their eyes blazed with Visine and fire. They told me that with Macintosh, they were going to "put a dent in the Universe." Their leader,

Steven P. Jobs, told them so. They also told me how Jobs referred to this new computer:

Insanely great.

Ten years later, I am boarding a Metroliner at New York City for quick overnight to Washington, D.C. In my left hand is a seven-pound gray box several times more powerful, but a thousand dollars less expensive, than the object I viewed in wonder that day in November. It is a PowerBook, the latest of my four Macintosh computers.

It is my typewriter, my communications center, my Rolodex, my Filofax, my alarm clock, my fax machine, my notebook, my database, my calculator, my file cabinet, and my opponent in chess and the slaughter of space aliens. It runs on a battery as big as a pack of baseball cards, though I'm just as happy plugging it into a wall socket. As the train pulls out of the station, I slip the PowerBook out of its case and press the space bar on its keyboard. A pleasant chime rings out, and the screen goes from a dusky fog to a familiar still life of little pictures on a lightly dotted whitish background. I have been using Macintosh for ten years now, and each time I turn it on, I am reminded of the first light I saw in Cupertino, 1983. It is exhilarating, like the first glimpse of green grass when entering a baseball stadium.

I have essentially accessed another world, the place where my information lives. It is a world that one enters without thinking of it . . . an ephemeral territory perched on the lip of math and firmament. Using the keyboard and mouse, one can reach into a metaphoric landscape, which has long become familiar. Though few know all the jargon identifying the peculiar Macintosh furniture—menu bars, title bars, elevators, close boxes, pull-downs and pop-ups—they become as cozy as the living room you grew up in. It's home. And in this place, you find familiar things. The paper you were working on. The spreadsheet figures you entered yesterday. Two different layouts you were considering for a publication you are designing. Even the simulated F-16 fighter jet you were piloting into a hostile zone near the Strait of Hormuz. This is a place with no physical substance, but it is of course wrong to assume that what happens there is in any way intangible. The work you perform there is real.

Very few tools transform their culture. Macintosh has been one of them. In the decade since the Mac's debut Apple has sold over twelve million Macintoshes—the sales rate of PowerBooks alone is over a million per annum. Extending the Macintosh style of handling information even more broadly are many millions more computers that run systems that owe just about everything to the Macintosh, notably Microsoft Windows.

The Macintosh has become a symbol of a sort of intellectual freedom, a signifier that someone has logged into the digital age. On television you see a Mac on Jerry Seinfeld's desk. It peers at you in the background of authors' photographs on book jackets. A newspaper reports breathlessly of producers conducting rapturous relationships with PowerBooks, of screenwriters sleeping with them. A magazine writes of a movie mogul who "grows rhapsodic" when he speaks of the device, and credits it for a career change and possibly even resolution of a mid-life crisis.

It took some time for people to see the light, but now it is everywhere, not only on personal computers but in television commercials that ape the look of its screen, and soon on cable television controllers and hand-held "personal communicators." The ideas of Macintosh no longer belong to the future: they dominate the present. And they will shape the way we cope with the future.

This book is about how technology, serendipity, passion, and magic combined to create what I believe is the most important consumer product in the last half of the twentieth century: the Macintosh computer. I will trace how Macintosh came into being, why it is so important, and how it already has set a process into motion that will eventually change our thinking about computers, our thinking about information, and even our thinking about thinking. In terms of our relationship with information, Macintosh changed everything.

I will also try to describe why, after a decade of using Macintosh, I still find it exhilarating.

I certainly don't claim that Macintosh is perfect. (At the time of its release it in some ways wasn't even adequate.) Certainly, I acknowledge that Macintosh is but a step in a path that was probably inevitable, the trail leading to a Digital Nirvana where all information, all music, all pictures, all voices, all transactions, and all mental activity gets parsed into seething bits of ones and zeros.

I am saying, however, that Macintosh was the crucial step, the turning point. Before 1984 the concept of ordinary human beings participating in digital worlds belonged to the arcane realm of data processing and science fiction. After Macintosh, these digital worlds began to weave themselves into the fabric of everyday life. Macintosh provided us with our first glimpse of where we fit into the future.

Though at the center of this story is a personal computer, sold by computer dealers in various forms over the last ten years, Macintosh is actually a creative expression of dozens of people, beginning with an idea first expressed in 1945. Humans often anthropomorphize the objects they use, especially when they become fond of their interaction with those objects. Almost everyone who comes into contact with Macintosh becomes enchanted by its personality. But by and large people seem to regard the emergence of this personality as a sort of random phenomenon, something that just happens once the computer leaves the factory and acclimates itself to its new surroundings.

Macintosh indeed has a distinctive demeanor, but this is a result of human effort and creativity—just as the traits of a character in a novel or film stem from the imagination of its author. Alan Kay, whose earlier breakthroughs in computer science and philosophy helped make Macintosh possible, has written:

> As with most media from which things are built, whether the thing is a cathedral, a bacterium, a sonnet, a fugue or a word processor, architecture dominates material. To understand clay is not to understand the pot. What a pot is all about can be appreciated better by understanding the creators and users of the pot and their need both to inform the material with their meaning and to extract meaning from the form.

Macintosh's creators viewed themselves as artists. Those who conceive of that term in the traditional manner—painters in smocks, poets in garrets, auteurs in film school—have to stretch a bit to snare this concept. The Mac creators are emblematic of a new kind of artist spawned by the protean nature of the computer.

Macintosh makes it clear that we are now witnessing a first flowering of a new form of expression, where architects of technology create interactive software that embodies their own, sometimes radical, visions. By using these products, we (most often unconsciously) experience those visions. They color our own thinking. We are transformed by them. Though the grammars, aesthetics, and even the jargon of this rather ephemeral art form have yet to be fixed, there is a quiet understanding among those working in the front lines of software design that they are participating in the most vital means of expression in our time.

During the Renaissance, a period frequently evoked by those working on or developing products for the Macintosh, painters undoubtedly agonized over the smallest details of their paintings. Every brush stroke told a story. In the early 1980s in Silicon Valley, furious disputes in aesthetics were waged over the likes of how many times an item on a drop-down menu should blink when a user dragged the cursor over it. (The Macintosh artists decided on three, but to appease those insisting on a lesser increment, they granted users the option to adjust the number.)

Keeping that in mind, it makes sense that in the process of documenting how Macintosh made its mark on the world, I should also explain how the creators of Macintosh made their mark on *it.* As it turns out, these artists were not bashful in helping me do this. I got that idea on the very first day I saw the machine, when I first met the Mac Team.

But first I should explain who I am, and how I wound up in Cupertino, California, on a sunny day in November 1983, staring in wonder at a machine that would change so many things.

My presence was anything but foredestined. For most of my life I had been a stranger to science and an uneasy companion to technology. If in the course of my education anyone had bothered to tell me about C. P. Snow's two cultures—science and the humanities, flopping against each other like grunting sumo wrestlers—I would have readily embraced the concept, rooting for the humanities all the way. It was not that I didn't enjoy science. I simply felt it had nothing to do with me.

In eleventh grade—1966—I listened compulsively to

Bob Dylan and the Paul Butterfield Blues Band and nearly flunked geometry. At Temple University, I played bad guitar in coffeehouses, took all the Shakespeare the place had to offer, and utilized a strange loophole in the academic rulebook to replace my two-semester science requirement with independent study in a topic of my choosing. I selected rock music. Instead of learning physics I wrote essays on the Rolling Stones and the Band. My final was a comparative study of the just-released Crosby, Stills, and Nash version of "Woodstock" and Joni Mitchell's original.

The computer science department was housed in Spellman Hall, one of several stupendously banal new buildings at Temple's North Philly campus. I think I went in there once, for a drink of water. Standing on the steps outside, dozens of students with white shirts and, sometimes, ties—ties!—conversed in what might have been Bantu, for all I knew. What I and my equally smug friends felt we *did* know was this: computers were evil.

Computers, we believed, had turned us into numbers. During biannual pre-registration, a hellish period where thousands of us jammed into a gym to secure courses for the semester, we were issued DO NOT BEND, SPINDLE OR MUTILATE cards, a handy symbol of the psychic slavery of attending college on a large campus. Computers tabulated our tuition increases, and mailed us the bills. Most damning of all, computers fueled the War Machine, that grinding, wheezing hunk of Kafka that murdered little babies and told us to report to 400 North Broad Street for a physical. Man, we hated that

War Machine. And it was so intertwined with those evil digit-crunching UNIVACS and IBMs, that the two were virtually synonymous.

Whatever went on in computer centers, it was bad, bad mojo.

I continued to retain my prejudices through the entire decade of the 1970s. So it was that when the biggest story of our era was breaking—an explosion of digital technology that would transform our civilization—I was elsewhere. I wrote stories about Bruce Springsteen, Doctor J, emergency ambulance squads, and denizens of cable television access channels, and tried not to think about the small but growing number of fellow writers who were abandoning their electric typewriters for something called word processors. (What a fat target *that* term presented!) The last thing I wanted was to have a television screen on my desk, with phosphorescent green letters shining in my eyes, making me blind and giving me cancer. I discussed this with my companion (now my bride), a fellow writer. We decided that, maybe, we would go as far as getting what certain trade periodicals then considered a more attractive alternative, an electronic typewriter. Several pioneering journalists had taken this step and then paid the bill by writing self-congratulatory accounts of their daring.

Were it not for a wildly fortuitous phone call in 1981, I might have wound up in their ranks, and thereafter consigned to a lifetime of technological cluelessness, stumbling from one wrong tool to the next—Epson QX-10, Osborne, Kaypro, Radio Shack TRS-80 . . . a computational trail of tears. There are people like this, doomed to buy these quickly orphaned husks of sand and plastic—digital losers, never quite catching the slipstream of our age. I could have been one of them, and probably would have been . . . had it not been for that telephone call in 1981, from Jane Fonda.

Actually, the call was not from Jane herself, but her production company. Would I be interested in doing a magazine story about a strange breed of human called computer hackers, and then selling the rights to the story back to them, so they could make a movie from it? Sure. I convinced *Rolling Stone* to assign the article to me, and I flew off to California, figuring if there was anything at all to this story, I would find it there. I was thirty years old and had never touched a computer.

Within hours after my arrival in the Golden State, I was stewing in a hot tub in the mountaintop retreat of Jim Warren, a gregarious Silicon Valley gadabout who was known for inventing the West Coast Computer Faire, an annual Wirehead Woodstock. Also in the tub were Tony Bove and Cheryl Rhodes, a pair of Grateful Deadheads who were sort of Jim's elves, living on his property while they edited a magazine about the possibilities of sending millions of pages of data by radio to people with personal computers. Maybe it was the temperature contrast—it was a soggy fifty degrees in the Bay Area though the tub was, well, hot—but more likely the conversation that made me violently vertiginous. Though I had tried to read up before my virginal exposure to the computer culture, I was quickly swamped by the discussion, which ranged from Trojan horses to something called CP/M. I had expected to be lost in jargon. What I did not expect was the excitement, and wonderment, with which all this jargon was voiced.

Slowly, I came to understand that a powerful, transforming force had been unleashed upon the world. So began my quest. Much as I immersed myself into Jim Warren's hot tub I took a thorough soak in the community of technoids. Though some of their character traits bordered on Dickensian, the most interesting thing about them was their vision: their goal was not primarily to make money (though a startling number of them were newly minted multimillionaires); they mostly did what they did because they loved doing it. But almost all of them believed that the fruits of their labors would empower ordinary people and perhaps even, in some subversive way, nudge the collective thought process, the group mind, toward the keystones of their philosophy, which were embodied in technology. These principles were decentralization, sharing, and the belief that knowledge was a good in and of itself. Amazingly, as personal computer technology was filtering into the *Weltanschauung,* their vision was actually filtering into the mainstream.

By late 1982, What the Nerds Had Wrought was imploding into the national mindset. It was at the end of that year that *Time* magazine would actually feature a machine, a personal computer, as the Man of the Year. *Time*'s cover was but the latest domino to fall in an avalanche of indicators that computers were the hottest thing going. No one was sure yet what the theme of the eighties would be, but it was quite clear that the Thing of the Eighties would be the personal computer, and that these objects would be with us for decades thereafter. As a journalist, I was at the epicenter of a historic tectonic shift.

I had stumbled onto The Big Story: the geeks were changing the world.

By late 1983 I was putting the finishing touches on a book called *Hackers.* Of course, I used a computer to write it, an Apple II. My first machine, it was a continual revelation, but in many ways it was frustrating to use. While the Apple II was a valuable tool, it bore in many ways its hobbyist roots. Not surprising since the industry itself was still in swaddling clothes. Word processing with a personal computer in the very early 1980s was like listening to a crystal set in the early days of radio—you could make out the broadcasts, but only by tinkering with the wires.

That era, however, was allegedly on the brink of extinction. The computer world was abuzz with rumors of two potentially earth-shattering computers. One was code-named Peanut, made by IBM, whose role in the industry was something like that of the Empire in the *Star Wars* series—dominant, invincible, and (to those who understood the Force) unspeakably vile. In 1981, IBM had belatedly entered the personal computer market with its PC, encountering as little resistance as Cortés had upon entering the Aztec capital. Though not particularly advanced in its technology, the IBM PC quickly became the standard. Now word had it that IBM had directed an army of programmers, in keeping with its "Human Wave" strategy of product development, to create a computer that would dazzle the legions of potential users who would devour the technology, if only they didn't fear and loathe it so much. I attended the introduction of this wunderkind, officially called the PC/Jr, and could not believe my eyes. It was a singularly bland morsel of technology, an intentionally crippled version of IBM's very successful PC. The Human Wave had generated barely a ripple.

Nowadays, the PC/Jr is a distant memory. Macintosh was another story. All over the Valley, people were whispering about how a small group of geniuses was devising something along the lines of Apple's impressive but prohibitively expensive Lisa computer. Introduced in January 1983, Lisa had been acclaimed as offering breakthrough technology, but few could afford it. Hopes abounded, however, that this new computer would break through to the masses, single-handedly launching the computer age into the stratosphere. Those depressed by the ease with which IBM had rocketed ahead of Apple looked to this new machine as the magic bullet that could stop Big Blue in its tracks. Very little in the way of specifics had leaked out of Apple, but it was common knowledge that the shipping date had slipped more than once. Were we in for another blizzard of groundless hype?

I confess that I hoped not. As an Apple II user, my heart was already with the Cupertino crowd. I had begged my editors at *Rolling Stone* to let me do a feature on this supposedly groundbreaking new machine and the young iconoclasts who had produced it. Since this would offer Apple advertising that one simply can not purchase, I expected the Silicon Valley version of a red carpet. To my surprise, my contact at Apple's PR firm, Regis McKenna (we journalists called these young women Regettes) informed me that access would not be forthcoming—unless *Rolling Stone* put the Macintosh team on its cover.

The odds of Jann Wenner agreeing to displace Sting in favor of a bunch of computer nerds were approximately one to googolplex.

Only one person in all of corporate America would have made such an absurd demand. This of course was Apple's chairman and the impresario of Macintosh, Steven Paul Jobs. His was a lifetime spent on the borderline between chutzpah and hubris. At twenty-eight years old, he was cofounder of a company that had quickly found itself in the Fortune 100. He was widely recognized as the symbol of American innovation and entrepreneurial cunning. Quite an achievement for a young man who only a decade previous was stumbling around India with a backpack, a spiritual hitchhiker without portfolio. His most noticeable trait was his charm, which he could seemingly turn on at will. Conversely, he was also legendary in his tactlessness, a character trait for which he had as yet not paid a price. He had over seven million shares in Apple stock (worth around a quarter of a billion dollars) to prove that he could get away with such behavior.

In this case, however, he backed down. Publicity was essential to nurture this tender enterprise in its first exposure to the marketplace, and the most saleable angle was the energetic team of Macintosh wizards. The only hitch was this: Macintosh still wasn't finished. After several deadlines set and unmet, January 24, less than eight weeks away, was the ultimate deadline, cast in granite. That was the annual stockholders meeting, at which the computer would not only be publicly introduced, but officially shipped. On January 25, when the frenzied hordes stormed the Apple dealerships, they would find Macintosh for sale. But only if the wizards finished it.

In light of this, my Regette begged me not to take advantage of the natural gregariousness of the Mac team. Though the ironclad policy was that every interview had to be chaperoned by a Regette, she knew that public relations people were held in disregard by the Mac team. If they decided that the *Rolling Stone* guy passed music, they'd even slip him their phone numbers for off-the-record conversations, no matter what their PR people said.

This is pretty much what happened. But Apple need not have worried. By the time sympathetic Mac Team members had, by some Cupertino equivalent of the Underground Railroad, helped me to lose my escort for hours of uncensored conversation, I was already a convert to Macintosh. This was partially due to a propagandizing routine that Apple had perfected to a T. But mostly, it was due to Macintosh.

. . .

Everyone called the building Bandley 3, signifying the third structure Apple had raised on Bandley Drive. This building was smack in the middle of the Apple "campus," which looked like and was a boring office park—prefab structures surrounded by asphalt. Parked on the lots were the best cars Japan had to offer. While the faux adobe facade of Bandley 3 was as exciting as Wonder Bread, a surprise awaited those who got beyond the reception area. A spacious lobby, complete with expensive mock skylights, awaited them. It held, among other things, a Bösendorfer grand piano, an arcade-version video game machine (Defender), a BMW motorcycle positioned like a heavy-metal work of art, and a Ping-Pong table. In a kitchen off the lobby was a refrigerator loaded with juices, soda, and various Calistoga waters. A compact disk player, rare for its day, fed Chopin and the Rolling Stones into fierce-looking, six-foot Martin-Logan speakers.

I hardly noted these wonderful toys before I was directed to the Matisse room. (Another conference room was called Picasso.) At this juncture, Barbara Koalkin, an Apple marketing manager, arrived, bearing the first Macintosh I would ever see. Enclosed in plastic casing of muddy beige, it was slightly bigger than a shoebox on end, about fourteen inches high. A small television screen covered its upper half.

After she turned the machine on, and I had sufficiently marveled at the quality of its display, she opened the file for the MacWrite word processor, and motioned for me to type a few words. I did. Then she did something unexpected. Using the mouse, she swiped the blinking line at the end of the sentence back over the words, which immediately turned white against a black background. Then, in one swoop of the mouse and a click, she did something that changed the words I had just typed. The sentence I had just written in a staid serif typeface suddenly was pushed leeward. It had become perfect *italic.*

I instantly recognized that every computer user was now potentially a publisher. I moved in to play with the machine some more, trying out things in MacWrite, and then a marvelous program called MacPaint. If I had not been somewhat familiar with the process of using personal computers, I would have thought I was seeing merely pleasant visual events and stunts. But for several years I had been struggling with the torture of mastering even simple tasks on the humble machine that was supposedly the most futuristic thing I owned. This did not seem like torture. It seemed like the future itself.

Poor Barbara Koalkin had to pry me away from the machine in order to give her canned spiel, which was sort of an overture before the opera, introducing themes I would hear developed with great intensity later on in the performance. I don't recall a word of it, really, but my notebook shows that I was dutifully jotting down key phrases, like "designed to be low-cost personal computer," "personal productivity tool for knowledge workers," and "we want everybody in the world using Mac software."

Anyway, I was dazzled, a feeling that would only accelerate as the day went on. Each person I met was a young wizard bubbling with enthusiasm—I could almost feel electricity crackling as they told me their stories. Jerry Manock, the industrial designer who had literally molded the Apple II and now the Macintosh. Mac would change the world, he said. Mike Boich, whose job title, formalized by his business card, was "Evangelist." It was his job to convince software developers to write programs for Macintosh. Mac, he confirmed, was going to change the world. Chris Espinosa, an original Apple employee, was now in charge of Macintosh publications. This computer, he insisted, will change the world.

Then I was off to another cubicle to see Joanna Hoffman, the first marketing person on the Macintosh team. Joanna totally threw me for a loop. After earning a physics degree at MIT, she had shifted to the humanities, attending graduate school at the University of Chicago, specializing in an obscure corner of Near Eastern archaeology. The overthrow of the Shah of Iran had closed off all the relevant digs; she was faced with a choice of focusing on another area or changing careers. It triggered a deep reevaluation of who she was, who she wanted to be. "I decided I'd been living in the past so long that I felt I now wanted to be in the future," she said. So naturally she went to California and wound up at Apple.

Joanna talked, in a voluptuous eastern European accent, about how Macintosh would be a global phenomenon, designed from the get-go to accommodate the quirks of other languages and cultures, from kanji to

Cyrillic. What I remember most about that encounter was that as Joanna spoke, her Macintosh was sitting on her desk, and *she could not keep her hands off it.* Every second sentence or so she would go back to it, caress it, stroke it, as if it were some rare breed of cat. It was eerie. I finally pointed out what she was doing, and she smiled sheepishly. "It's such a cute little beast," she said.

Then she told an anecdote that would turn out to have prophetic resonance. She was showing the program to some Italian businessmen. At first they were skeptical, but then Joanna cranked up MacPaint and they went crazy. "We couldn't get them out of the room," she said. Then she suddenly got very serious.

"You know, it's hard to tell people that something elegant and airy is powerful," she said. "I tried to stress the applications, emphasize that this can be a very serious machine. They wanted to paint. True, there's no need to make a computer burdensome. It can be delightful—yet very powerful and useful."

Before I could chew on that, I was led into the inner sanctum, the room where the engineers were actually trying to get the Macintosh out the door. I was to have lunch with "Bill and Andy," neither of whom I had met. But my curiosity had been whetted—almost every person I had interviewed said, "Wait till you meet Bill and Andy," as if this would certify the special nature of the Macintosh experience.

I met Bill Atkinson first. A tall fellow with unruly hair, a Pancho Villa mustache, and blazing blue eyes, he had the unnerving intensity of Bruce Dern in one of his turns as an unhinged Vietnam vet. Like everyone else in the room, he wore jeans and a T-shirt. "Do you want to see a bug?" he asked me. He pulled me in his cubicle and pointed to his Macintosh. Filling the screen was an incredibly detailed drawing of an insect. It was beautiful, something that you might see on an expensive workstation in a research lab, but not on a personal computer. Atkinson laughed at his joke, then got very serious, talking in an intense near-whisper that gave his words a reverential weight.

"The barrier between words and pictures is broken," he said. "Until now, the world of art has been a sacred club. Like fine china. Now, it's for daily use. We're going to make it so easy to be creative that people will have no excuse not to confront their own artistic ability."

We were met for lunch by Andy Hertzfeld: late twenties, compact, elfish, bespectacled, and overflowing with energy. His business card read "Software Wizard." His words tumbled out with the mile-a-minute cadence of a small boy describing a demolition derby.

"There are two barriers that keep one hundred fifty million people from using the computer," he said, as soon as we settled into a booth at a suburban saloon. "First, it's too expensive. Second it's too hard to use. You have to get immersed in this muck of horrible stuff. Computers are great, but they don't do any good if it doesn't reach the common man. But we're bringing computers to the people for the first time. When we designed Macintosh, we aimed at ourselves—people. Like when I got my first stereo, I knew I wanted it in my life. We want the man on the street to get Mac and feel that incredible potential . . . when every person has a computer he or she can relate to, it's going to change the world!"

Our food had arrived, but no one had touched it. "We're all maniacs," Andy said. "People want to be computer scientists, but we are also hackers, trying to make Macintosh incredibly small and tight and fast. The thing that turns me on is making this great computer." Then he stated the obvious. "I get emotional about my work."

Atkinson nodded. "You're doing this because that's the dream," he said. "Don't mess with my dream, and I'll like you."

By that time, all I could think of was the moment in *Butch Cassidy and the Sundance Kid* when the protagonists looked at each other and asked, of their superhuman pursuers, Who *are* these guys?

And then there was Steve Jobs. It was late in the day, and I was talking to some people in the Bandley 3 lobby. He appeared out of nowhere, trim and handsome in a navy sweater (no shirt) and jeans. In lieu of a traditional greeting, his first words to me were, "I think you're making a big mistake by not putting the Mac on the cover." His eyes bore down on me out of a somewhat hawklike face, and I immediately became flustered.

This was my first exposure to what Jobs's subordinates would call "the reality distortion field." Though I was in no danger of accepting his premise, the effect of his tirade was impressive. The people standing around us looked embarrassed. Then, just as suddenly, Jobs took off, headed to douse some unidentified conflagration. It was understood that we'd have dinner that night.

Sometime after the appointed hour, we got into his car and headed for a nearby pizza house. He immediately picked up the thread of his previous complaint, lobbying me for a cover shot. (As if I could do something about it.) It was Jobs's contention that *Rolling Stone* was on the ropes, running crummy articles, looking desperately for new topics and new audiences. The Mac could be its salvation!

I asked him if he'd even read the magazine lately, challenging him to cite any of these crummy articles. Well, yes, he replied. Just the other day on the airplane he'd read a copy of the *Stone,* and thought that the cover story, the one about MTV, was really awful. And began to tell me what a piece of shit that was.

I told him I was the author of that particular article. To his credit, he didn't flinch, or attempt to mitigate his comments. He did, however, curtail his withering critique. We downshifted the conversation to small talk. When we got to the restaurant, there was an awkward silence as I took out my tape recorder. Though Jobs knew the Macintosh required an unprecedented amount of publicity, he had a deep mistrust of the press. A year before, he had been traumatized by a story in *Time* that he considered a cruel personal attack by a reporter he once trusted. "I know what it's like to have your private life painted in the worst possible light in front of a lot of people," he said. "It was a hatchet job." When I told him I didn't think that article was that devastating, he resisted, noting that the point was not how others viewed the article, but how *he* did. Ever since that perceived betrayal, he had been leery of my ilk.

Yet once I switched on my recorder, Jobs became enthusiastic and candid. "Computers and society are out on a first date in this decade and for some crazy reason we're just in the right place at the right time to make that romance blossom," he said. And then he told me about Macintosh.

"I look at most of the people I get to work with as artists. I look at myself as an artist if anything."

"Really?"

"Sort of a trapeze artist," he joked.

"With or without a net?"

"Without." Then he turned serious again. "It's a way of expressing feelings. Wanting to put something back into the world." He pondered his own words for a moment. Then, with great animation, he said, "You know, we don't grow most of the food we eat. We wear clothes other people make. We speak a language that other people developed. We use a mathematics that other people evolved . . . I mean, we're constantly *taking* things. It's a wonderful, ecstatic feeling to create something that puts it back in the pool of human experience and knowledge. I think actually one can influence things as much or more from the private sector than the public sector. I'm one of those people who think that Thomas Edison and the light bulb changed the world a lot more than Karl Marx ever did. And we have this incredible chance to do that in the next five years.

"I don't want to sound arrogant but I know this thing is going to be the next great milestone in this industry.

If it's not, I'll just go back to Tibet or something. Retire from this material life. Every bone in my body says it's going to be great. And people are just going to realize that and buy it."

Later he modified the superlative, using the phrase he had been hammering upon the Macintosh group for over a year—not just great, he said, but *insanely* great.

2

In 1945 Vannevar Bush, a former vice president of MIT and then the director of the country's Office of Scientific Research and Development, wrote an essay in the *Atlantic* entitled "As We May Think." It was the end of World War II. Bush had been instrumental in channeling the efforts of thousands of scientists to produce techniques and devices of massive destruction and cold-blooded mayhem. Thankfully, this sort of thing was over, he thought. Now the question was, as he posed it, "What are the scientists to do next?"

His answer sparked a chain reaction that led, almost forty years after the article was published, to the Macintosh computer. Bush contended that the major scientific and engineering effort in postwar America should be the transformation of the way we process, retain, and retrieve information.

Obviously, he was thinking about the electronic computer, a monstrous number-crunching mechanism just being developed at the time he wrote. Bush understood, as did very few in those days, that the underlying technologies of this new tool were not limited to a more efficient duplication of the labors of roomsful of human calculators. He envisioned these machines as processing symbols as well. Bush called for the development of a new sort of language, one capable of sucking in many kinds of input—mathematical, textual, vocal, and visual. These could later be displayed by devices like cathode-ray tubes, "dry photography" (essentially, the technology we use in copying and fax machines), and new, highly compressed forms that he called "microphotography." "The Encyclopaedia Britannica could be reduced to the size of a matchbox," he predicted. One could gather the entire written output of the human race and load it into a single moving van.

This same lingua franca—which we now readily know as digital format—would be put to use in creating original documents, Vannevar Bush said in 1945. And then he described what would later be known as word processing.

If this were not remarkable enough, Bush went on to give the design specs for his dream machine, one he called the memex. This "sort of mechanized private file and library" would put to use the vast knowledge stuffed into the aforementioned moving van—allowing the memex user to produce a new document that seamlessly integrated the vast sprawl of the human legacy into a book, letter, audio recording, or causal notation.

What does a memex look like? Bush explained:

> It consists of a desk, and while it can presumably be operated from a distance, it is primarily the piece of

> furniture at which [the user] works. On the top are slanting translucent screens, on which material can be projected for convenient reading. There is a keyboard, and sets of buttons and levers. Otherwise it looks like an ordinary desk.

Stored within this desk is information, tons of it. "Books of all sorts, pictures, current periodicals, newspapers . . ." So vast is the storage space that Bush estimated that even if the user jammed five thousand pages of material daily into the desk, he or she would never live long enough to fill the data bucket. Bush had the right idea here—access to reams of information—but he didn't grasp how much easier it would become to store all materials, the whole moving van of human history, in a central location, available to the memex user by a wire umbilical. Where Bush's vision turned deadly accurate was when he outlined how the memex user would find and retrieve what he or she wanted. Besides a standard keyboard, the memex would have rows of buttons and levers. By pushing the buttons and pulling on the levers, a memex jockey would navigate through the information, calling up files, shuffling it to one of the screens, stepping through the text like a frenzied Evelyn Wood student, ultimately locating the desired penetrating thought or factoid. Bush was inventing what we now call "information surfing," where one rides the crest of a concept and, flicking the board in a perfect 360, catches a new concept. Sitting in the cockpit of the memex like a fighter jockey, Bush's text pilot cruises through a realm that Bush obviously envisioned quite clearly, but would not be named for two human generations: cyberspace. The tools for navigating this terrain at the speed of imagination would be the buttons and levers of the memex machine.

"Presumably," wrote the nation's chief scientist, "man's spirit should be elevated if he can better review his shady past and analyze more completely and objectively his present problems." The ability to manipulate information, he implied, could be our very salvation.

The issue of the *Atlantic* with Bush's essay eventually found its way to a bamboo structure on stilts in the Philippine island of Leyte, a hut converted to a library for naval personnel by the Red Cross. Hiroshima and Nagasaki were in deadly smoke. And a twenty-year-old naval radar technician named Douglas C. Engelbart was idly waiting for a boat to take him home. One day he went to the library and read "As We May Think." It would not be too much of an exaggeration to say that on that day in 1945, the seed was planted that would one day bloom into Macintosh.

The germination, however, was slow. Engelbart didn't think much about Bush's vision for five years, when he had become a fledging engineer for the agency that was destined to become NASA. He was doing research with wind tunnels; this, in 1950, was a hot job. He had just proposed marriage to a woman he had met at work, and he felt his life stretch out before him. As lives go, it looked pretty good. Yet there was something, he felt, terribly lacking. "Suddenly," he would later recall, "it all seemed much too flat to accept."

Douglas C. Engelbart was constitutionally a dreamer.

And for the next few weeks, he attempted to dream of a new career for himself, something—this sounds so idealistic that it's almost wacky—that could actually help shape the future of humanity. He decided that instead of solving a particular problem, he would influence the very act of solving problems. He would give the world a tool to improve its abilities to wrestle problems into submission. Something to *augment* human powers. That was the word he used, augment. The other word he would come to use was *crusade.* Engelbart was embarking on a crusade to augment human capabilities by applying new technologies and developing ways to interact with that technology. He ultimately would realize, and even surpass, what Vannevar Bush had written in his terribly important yet unappreciated essay in the *Atlantic.*

Crusades are not completed in a day, even in a year. Engelbart quit his job, moved to Berkeley with his new bride, and earned a Ph.D. in the budding field of computer science. He was hired by a think tank called the Stanford Research Institute (SRI), and he set up a group that he called the Augmentation Research Center. He cajoled a small grant from the government, and set out to change the way the world worked.

And did.

I once went to visit Doug Engelbart. His employer in 1983, the inheritor of what was left of the Augmentation Research Center, was a phone networking company called Tymshare. Amazingly, the building that housed Engelbart was in Cupertino, a quick turn off Bandley

Drive. Unbeknownst to either of us as we spoke, less than two blocks away, Apple Computer's Mac Team was working feverishly to realize some of Engelbart's dreams (though they were hardly aware of the man himself).

The Tymshare building was standard-issue Silicon Valley architecture—featureless, clean, and low to the ground. Just another chip on the dense circuit board of Santa Clara County. Meeting me in the reception area was a trim, avuncular man with gray hair and a well-trimmed beard. He was in shirtsleeves. His greeting was warm yet understated. Douglas C. Engelbart led me to his office, a cubicle in one corner of a sprawling room filled with file cabinets and similar warrens. The Moses of computers did not even rate an enclosed office.

I later read in Howard Rheingold's book *Tools for Thought* what a friend had written of Engelbart: "When he smiles, his face is wistful and boyish, but once the energy of his forward motion is halted and he stops to ponder, his pale blue eyes seem to express sadness and loneliness."

Indeed, there was something sad about our visit. I got the impression that Engelbart didn't have all that many visitors. In recent years, partially as a result of the revolution that the Macintosh has spurred, Engelbart has finally won considerable recognition, if not riches. But when I visited him, Douglas C. Engelbart was still a name that would draw a blank stare, even in the techno-gossip hotbox of Silicon Valley.

Unjustified obscurity was something that Engelbart had learned, though not happily, to live with. In 1962, writing to his virtual mentor Vannevar Bush for permission to reprint some of the *Atlantic* article, he complained, "I had . . . almost nothing but negative reaction from people [before working at SRI] and for several years here, too." He had hoped things would change in the early 1960s when he began to publish; he told Bush that he was preparing "more than just a report to a government agency. To me it is the public debut of a dream, and the overdue birth attests to my emotional involvement." In 1963, thirteen years after first adopting Vannevar Bush's vision as his launching pad for the modernization of man, Engelbart published a paper called "A Conceptual Framework for the Augmentation of Man's Intellect." Like Bush before him, Engelbart complained that the accumulated knowledge of humanity had exceeded our ability to handle it. Only by "augmenting man's intellect" could we remedy this situation—resulting in better comprehension of problems, quicker solutions to those problems, and the conquest of previously insoluble problems. And here was the news: *The tools to perform this task were at hand. Here is how we will work in the future.*

And then Doug Engelbart predicted systems in which individuals would have personal computer workstations, networked to each other, and would even compose documents in which "trial drafts can be rapidly composed from rearranged excerpts of old drafts . . . you can integrate your new ideas more easily and thus harness your creativity more continuously."

Yes. In 1963, when computers cost hundreds of thousands of dollars, and when those few people who worked with computers generally did so by submitting stacks of punched cards to authorized tenders, here was a man who for years had envisioned word processing and networking. (Almost parenthetically, Engelbart explained how on-line dictionaries could be part of the writing process.) But that was only part of his vision, which involved a unified scheme to manipulate information and enhance creativity, a system that could dwarf "the combined effects of the printing press and the industrial revolution."

"That paper was the first time or place I came out with those thoughts," Engelbart told me twenty years later. "There was a lot of risk there."

And the response?

He looked at me, his blue eyes unflinching. "There was one little review someplace that said, 'Here's a description of a little documentation system.' "

Yet he pressed on—this was a crusade, after all—and his group, working out of World War II–vintage barrackslike huts on the SRI campus, pushed what was then the limits of computing. It was there that Engelbart invented windows.

Windows. For many of us, the Macintosh was a first exposure to windows, that technique of putting multiple views on the same computer screen. It's instantly clear how the metaphor applies: the screen is flexibly partitioned into any number of rectangles, each of which could be considered a window into a separate monitor, showing a display of a separate file, or even several rearrangements of the information in a single file. It was a way to realize a feature of Vannevar Bush's memex—those multiple cathode-ray tubes built into the desk-sized structure. Instead of sitting in front of several monitors, like rent-a-cops at a condominium security control room, we would require but a single screen. To be sure, what was most impressive about that first exposure was the simple fact that more than one file could be displayed simultaneously—quite an advance from the one-at-a-time regimen most of us were used to, pre-Macintosh.

Windows are really quite profound. Using them implicitly reshapes our relationship to information itself. Information is what we see when we look through those windows—a digital peep show where we flick open the shutters to information.

As Howard Rheingold would note, "The territory you see through the augmented window in your new vehicle is not the normal landscape of plains and trees and oceans, but an *informationscape* in which the features are words, numbers, graphs, images, concepts, paragraphs, arguments, relationships, formulas, diagrams, proofs, bodies of literature and schools of criticism." We now have a term for this informationscape: cyberspace.

We received the term courtesy of William Gibson, a science fiction writer. In his *Neuromancer,* the high riders of computer networks "jacked into" it. . . . a "consensual illusion." But there is a weird provenance here—unlike other prescient sci-fi writers who give us terms for things that haven't been invented yet, Gibson's winning phrase, putatively used by those living in a near-future dystopia, actually applies to Engelbart's informationscape and its current descendants. In the age of computers, the best science fiction writers are no longer speculative prophets, but interpreters of a newly synthesized reality, constructed of mathematics and information.

Thus it appears that instead of inventing cyberspace, Gibson identified it. It had been there, we now understand, for years. Cyberspace, says essayist and lyricist John Perry Barlow, is where conversations are conducted when two people talk on the telephone. But most often it is associated with a landscape of data. Or as Michael Benedikt, an architecture professor, described C-space, "The tablet become a page become a screen become a world, a virtual world. Its depths increase with every image and word or number, with every addition, every contribution, of fact or thought. Its corridors form wherever electricity runs with intelligence. Its chambers bloom wherever data gathers and is stored. . . ."

It was Engelbart who devised the first tools to propel one through cyberspace. Best known is the mouse. An odd companion to a keyboard—because its purpose was something that could not possibly have been envisioned until Engelbart constructed his system—it would be used to reach into and manipulate a world constructed only of information. The requirement was that this task had to be so smoothly integrated into the overall system that people would not even realize that they had jammed their hand into the Phantom Zone. It had to feel like the mouse-handlers were actually working on real paper.

Before Engelbart's group settled on the mouse, it ran through all sorts of weird-science accouterments: joysticks, track balls, light pens, and a device that Irwin

Corey would have loved—a sort of steering wheel controlled by one's knee. (At one point, Engelbart attempted to test the *worst* imaginable means of manual input, and published a paper on the matter entitled "Experimental Results of Tying a Brick to a Pencil to De-Augment the Individual.") Engelbart ultimately described the traits of the mouse that elevated it over these alternatives:

> [It is] palm-filling size, has a flexible cord, and is operated by moving it over a suitable hard surface that has no other function than to generate the proper mixture of rolling and sliding motions. . . .
>
> That the mouse beat out its competitors . . . seemed to be based on small factors: it stays put when your hand leaves it to do something else (type or move a paper) and reaccessing proves quick and free from fumbling. Also, it allows you to shift your posture easily, which is important during long work sessions . . . And it doesn't require a special and hard-to-move work surface . . . A practiced, intently involved worker can be observed using his mouse effectively when its movement area is littered with an amazing assortment of paper, pens, and coffee cups, somehow running right over some of it and working around the rest.

(I can attest, from great experience, to the latter claim.)

The first prototypes of the mouse were carved from wood, rounded blocks with little tracking wheels (potentiometers) underneath. They had three buttons, and if one looked at them a certain way, the buttons seemed analogous to a mousy nose and two tiny mouse ears. This, and the taillike cord trailing out of the device's rear, earned it the nickname that, quite by inertia, became its permanent appellation.

The name might have been arrived at whimsically, but nothing else about the mouse was an accident. Bill English, who at the time was one of Engelbart's uncaped crusaders, once explained how the mouse was crafted for comfort and control. "You want it fairly low, without your wrist on the table. Ideally, you'd like to use the same muscles you use to control a pencil."

In 1968, Engelbart unveiled his entire system to an astonished fall Joint Computer Conference in San Francisco's Civic Center. With his keyboard, his keypad, and his mouse, Engelbart embarked on a journey through information itself. As windows open and shut, and their contents reshuffled, the audience stared into the maw of cyberspace. Engelbart, with a no-hands mike, talked them through, a calming voice from Mission Control as the truly final frontier whizzed before their eyes.

It was the mother of all demos. Engelbart's support staff was as elaborate as one would find at a modern Grateful Dead concert. The viewers saw a projection of Doug Engelbart's face, with the text of the screen superimposed on it. At one point, control of the system was passed off, like some digital football, to the Augmentation team at SRI, forty miles down the peninsula. Amazingly, nothing went wrong. Not only was the future explained, it was *there,* as Engelbart piloted through cyberspace at hyperspeed. Now, so many people claim to have been at the demonstration that it's sort of a modern version of Babe Ruth's called home run. At the time, though, word of the performance did not spread to the world at large, to whom the revolution was directed.

Engelbart's project had a single major patron: the Advanced Research Project Agency of the United States Department of Defense. Unbeknownst to Timothy Leary when he attempted to levitate the evil Pentagon in 1967, this little-known branch of Defense was quietly kick-starting the computer revolution that would result in the Macintosh. I don't want to pretty this up too much now: the interest of the Defense Department was of course the development of systems that could toast the flesh of opposing soldiers and noisome bystanders.

But from the start, ARPA's leadership was enlightened: the very first person in charge of the Information Processing Techniques Office was J. C. R. Licklider, whose dreams and desires ran parallel to Engelbart's: to overthrow the status quo where computers were accessed only indirectly, by way of submitting punched cards and waiting, sometimes for days, for the results. Licklider was an early advocate of computer interactivity. In 1960, he was writing papers proposing a man-computer symbiosis in which "human brains and computing machines will be coupled together very tightly, and . . . the resulting partnership will think as no human being has ever thought and process data in a way not approached by the information-handling machines we know today." When he went to the Pentagon in 1962, Licklider, and later his like-minded successors, set up a system that allowed them to toss millions of dollars at research centers trying to realize this goal. For eight years the money flowed—modestly by defense standards, topping out at about $25 million a year—until some persnickety senators forced the agency to limit its spending only to projects with specific military applications.

Yet during its golden run in the 1960s, ARPA was probably the greatest government investment since the Louisiana Purchase. Compare it to its contemporary, the space program. The latter focused on a single mind-blowing goal, a moon landing, which was successfully met. And then the enterprise fizzled, becoming decreasingly relevant to the general public. The main benefits of the whole enterprise seem to have been Teflon, Tang, and a stack of very cool photographs. ARPA—by using its relatively meager bankroll (millions, not billions) to seed an entire culture devoted to transforming computers into instruments of communications and mental augmentation—bootstrapped a revolution that would change the way all of us worked, created, and thought.

Contemplate the actual products: Word processing. Personal computing. Desktop publishing. Spreadsheets.

Not that Doug Engelbart would personally reap the fruits of this transformation. Even as the trajectory of his thought kept rising in the early seventies, the clock was ticking on his pet project. In 1973, enchanted by the ideas of economist Peter Drucker, he coauthored a paper called "The Augmented Knowledge Workshop," which was based on the idea, formulated by Drucker, that information was destined to be the fulcrum of the economy. "By 1960," wrote Engelbart, "the largest single group [of Americans] was professional, managerial, and technical—that is knowledge workers. By 1975–1980 this group will embrace the majority of Americans." It was these knowledge workers who would sit in the cybercockpits of the Engelbart augmentation scheme: as he termed it, "the office of the future."

Knowledge workers. The office of the future. These two catch phrases would later be appropriated by the marketers charged with selling the Macintosh. But Engelbart's bosses at SRI weren't concerned with marketing his product. They tolerated him, as long as he was funded. And then he lost his funding.

In retrospect, it seems hard to fathom. It's as if someone, the biggest idiot in the annals of technology, had handed Edison his walking papers. *Lock up the lab, Tom! Take this junk with you!* Some people attributed the problem to what was increasingly becoming a mismatch between the conservative SRI and the iconoclastic Augmentation group. Others claimed that Engelbart's vision was too rigorous, that it demanded too much from the poor knowledge workers. (Hey, the future isn't easy.) In the early 1970s some key members of his team fled to a fledging computer research center set up by the Xerox Corporation. And in 1975, after a less-extravagant ARPA pulled its golden plug, the Augmentation Research team broke up for good. SRI sold the system to Tymshare. By then the "team" consisted of one: Douglas C. Engelbart, once again a solitary dreamer.

Engelbart's voice was brittle when he explained the move. "There was a slightly less than universal perception of our value at SRI." Not that Tymshare, his current employer, had a deep appreciation for what they had purchased. "It was a cheap way to take a chance," said Engelbart, not bothering to cloak his bitterness.

By the time I wound up in Engelbart's pathetic cubicle in 1983, his creation, now dubbed "Augment," was one of several office-automation systems Tymshare offered. There was little fanfare. Yet to Engelbart it was still his baby. He talked as if his system, not the evolutions of it like the Lisa and the upcoming Macintosh, was going to take over the world. It was logical to him that it should. *Here,* he seemed to say, *just watch.*

I sat there slack-jawed as he demonstrated. The most memorable portion of his system, perhaps because of its unfamiliarity, was the little one-handed "keyset" that could be chorded like an electronic piano. Using that in one hand, and a mouse in the other, he could make lines of text appear on the screen without once touching the standard keyboard. Seeing this, I had to admit that anyone who learned how to play this thing would have a huge advantage over the laggards wed to our current antique. But to get this part of his system in the mainstream, Engelbart would have had to overthrow a technology entrenched more doggedly than the Maginot Line—the QWERTY typewriter keyboard.

Everyone knows what a dog *that* interface is. If it seems like the QWERTY keyboard was laid out deliberately to slow down speed typists, that's because it was. If people typed too fast, they would overwhelm the machine—the keys would jam. So inefficiency was built in. A good idea for the nineteenth century, when text was produced by pounding a lever to make an impact on a piece of paper, but not so good a hundred years later, when hitting a key sends electrical impulses to a silicon chip.

Yet we stick with the interface; there's too much invested in the standard to abandon it. My own high school instruction in typing was nightmarish; so fumble fingered was I that after my mistakes were deducted from my word totals, my scores on the speed drills were usually gauged in negative numbers. Yet I knew, since my handwriting was universally regarded as illegible, that in order to make my thoughts known in the world, I had to learn the keyboard, and eventually I internalized the skills. This ability is so hard-won that I cannot imagine attempting something as elaborate as a chording handset.

Engelbart's system could take you farther, faster, than anything that came before it and arguably has not been eclipsed to this day. But it required several weeks of intensive training to master it. This is not a lot to ask of rocket scientists, brain surgeons, and plumbers—but, for knowledge workers this was and is an insurmountable barrier. Better to go slow every day for the next decade than to lose a few weeks learning to move at hyperspeed.

Still, Engelbart got farther than anyone had reason to expect by postulating an addition to the keyboard—the mouse—and seeing it accepted in his lifetime. To boot, he was the first to implement windows in computer screens, transforming a single video monitor to the multiscreened hydra envisioned by Bush with his memex machine. Since gathering the wealth of Croesus was not

Engelbart's goal, one might reasonably assume that his achievements brought him satisfaction.

When I suggested as much, Engelbart curtly gestured to his system, a full-fledged wonder with not only mouse and windows, but the chording handset, the networking, and the implied connection to the world's knowledge via the portals of cyberspace. He wanted *this* system, *his* system, everywhere. But he had no control over the future. His vision was at the mercy of those he inspired.

3

The next leap toward Macintosh would originate only a few miles from Engelbart's lab—a research and development arm of the Xerox Corporation called the Palo Alto Research Center, but known to computerheads everywhere as PARC. It would become famous, but not quite in the way its parent company intended.

PARC's establishment in 1970 came as a consequence of the Xerox Corporation's extraordinary success. Xerox was literally synonymous with copying machines; the revenues piled in and the profits piled up. But the company's future was far from assured. Making copies of documents was but one component in managing information. The emergence of advanced computing tools would create new components, possibly some much more significant than even the most advanced copiers. Xerox had gotten into the computer field by buying a California company called Scientific Data Systems, but Xerox's CEO, Peter McColough, realized that the "office of the future," to use Engelbart's term, was yet to be invented. He decided to spend some of Xerox's profits to invent it. Thus PARC was born.

Though the head of the research center was a respected scientist named George Pake, the key figure at PARC was a nonengineer, Bob Taylor. Straight out of ARPA, where he had succeeded J. C. R. Licklider as head of the Information Processing desk, Taylor ran his domain as if it were somehow a continuation of the ARPA effort to push computation into the realm of the intimate. As surely as an ACC basketball coach is familiar with the country's top high school prospects, Taylor had kept track of the nation's bright young computer scientists, particularly the freewheeling wizards who aimed to hack up miracles. And he landed the best.

It did not require much wooing. Word was out that something special was afoot in PARC's faux Aztec complex, built into a mound of rugged terrain on Coyote Hill Road near Route 280. The ARPA A-team gladly fled their universities and artificial intelligence labs to make a bit of history. (The pay wasn't bad, either.)

"This is really a frightening group of people, by far the best I know of as far as talent and creativity," said Alan Kay, one of the center's principal scientists, in 1972, smack in the middle of PARC's Golden Days. "The people here all have track records and are used to dealing lightning with both hands." Kay, himself a frighteningly talented person, also once claimed that "out of the one hundred best computer scientists in the country, seventy-six of them were at PARC." (One of his former colleagues, upon hearing this, remarked dryly that Kay's estimate was suspect, "for the very primitive reason that we didn't even have seventy-six scientists.")

Yet PARC's brainpower was truly impressive, and is now safely ensconced in legend. There still is a Xerox Palo Alto Research Center, in the very same building on Coyote Hill Road, but when people talk of PARC they generally refer to the collection of computer science magi who populated the center in the 1970s. Nerds now think of it as Camelot. While the nation discoed, PARC redefined computing. Its scientists sat on beanbag chairs (stacked to the ceiling in meeting rooms—at conferences people would file in, grab one, and flop down on the floor) and rethought the symbiosis between man and machine. And then they created a marvel. It was unique for its time, but now would be identified as sort of looking something like a Macintosh. But not as good as a Mac.

I missed visiting PARC during its Camelot era, but if I'd gone, surely the first person I would have sought out would have been Alan Kay. His intelligence is well documented: a radio quiz kid at age ten, a restless, unchallenged student through high school; a passionate musician distinguished at the pipe organ; and, as he discovered during his stint with the United States Air Force, a virtuoso computer programmer. He used his coding skills to finance his undergraduate education at the University of Colorado, and somehow came to the attention of Dave Evans, the head of computer science at the University of Utah. In 1966, Kay was admitted to Utah's graduate program.

Though these days the University of Utah is best known for its attempt to solve the world's energy problems through the magic of cold fusion, in the mid-1960s it was a hotbed of computer innovation. The hidden hand in bringing this about was, of course, ARPA. It had given Evans, a former Berkeley computer scientist, a five-million-dollar grant, which allowed him to go wild with computer interactivity and graphics. Evans pried a lot of talent from vaunted computer palaces such as MIT, but the biggest prize of all was Ivan Sutherland.

Sutherland had achieved canonical status in the field by devising a computer program called Sketchpad. He had concocted this as his MIT doctoral thesis in the early sixties, working on the TX-2, one of the first computers with a visual display, albeit an extremely crude one. "It filled a room," Sutherland later recalled. "It was about a twentieth the power of a Macintosh II, but it was at that time a very big machine." The TX-2 had a remarkable feature: using a flashlightlike wand called a light pen, one could input shapes into it. Sutherland had, since childhood, harbored a fascination for geometry and mechanical drawing (his father was a civil engineer). So, as he later explained, "it seemed like the most natural thing to make drawings with it."

Natural for Sutherland, perhaps. But few had imagined that this rough beast of a calculating engine could be transmogrified into a sophisticated system to create shapes, pictures, and blueprints. And when you created your shapes you could copy them, alter them, or store them. In 1977, Ted Nelson (whom we will meet when our story turns to HyperCard) gushed about Sutherland's wonder—"The Most Important Computer Program Ever Written," he called it—in his book *The Home Computer Revolution.*

> . . . working on a screen you could try out things you couldn't try out as a draftsman on paper. You were concerning yourself with an abstracted version of the drafting problem; you didn't have to sharpen any pencils, or prepare a sheet to draw on, or use a T-square or an eraser. All these functions were built into the program in ways that you could use through the flick of a switch or the pointing of the light pen. And the drawing itself existed in an abstracted version, that could be freely changed around with no loss of detail.

Sketchpad was not only the first drawing program, but was arguably the best, absurdly ahead of its time. It's as if the designer of the very first automobile had created a 1967 Corvette. Sketchpad was also, for many years, the thing that most clearly anticipated Macintosh.

Sketchpad did not *feel* like a computer program, at least none that had ever been thought of as such. It felt like . . . pictures. Like geometry. Like cyberspace. Suddenly, we could see the pictures of purely mental terrain that entranced Plato when he talked about mathematics. And we could play in it, wander through it together, point to things. Sketchpad was never destined for general distribution—it ran only on the TX-2, a computer with a production run of one, and was never ported anywhere else—and it never affected the culture nearly as much as less significant but flashier advances in computer science (like, say, chess-playing programs).

Still, when presented to a receptive mind, Sketchpad could twist that mind into a pretzel. Using the saurian machines of the fifties and sixties, it was all too easy to think of computers as simply crunchers of numbers, of shufflers of ones and zeros. Very few people, even among the new cognoscenti of computer scientists, had the imagination to see how versatile ones and zeros could become: they could be pictures, sounds, characters, the cells of new creatures, the building blocks of new worlds. Sketchpad changed all this; the program profoundly affected those who spent time with it. Among those was Alan Kay.

When Kay arrived at Utah, Dave Evans immediately dumped Sutherland's Sketchpad thesis on him. Kay understood its significance very quickly: Sketchpad made the computer an extension of a single person's mental terrain—made it personal. For the next two years Alan Kay attempted to design a new kind of hardware, something that most people thought a wretched excess: a personal computer.

He called it the FLEX. It was groundbreaking (or would have been, had it been built), because it wove the lessons of Engelbart and Sutherland into a compact package. "It had a tablet as a pointing device, a high-resolution display for text and animated graphics and multiple windows," Kay later described it. But its Achilles' heel was the interface, which its creator admitted in retrospect was "repellent" to users. One thing he got right was the use of little pictures, later called icons, which people could point to in the course of recalling previous work. But generally FLEX was a mishmash. "The combination of ingredients didn't gel. It was like trying to bake a pie from random ingredients in a kitchen: baloney instead of apples, ground-up Cheerios instead of flour, etc," wrote Kay.

This failure led Kay to an examination of what a "user interface" meant. The term commonly referred to a set of screen prompts and commands that allows a person to communicate his or her wishes to the computer. "The practice of user interface design has been around at least since humans invented tools," Kay later noted. Yet very little thinking had been devoted to promoting friendly, intuitive computer interfaces.

Could an interface be designed so that ordinary people could use it? This was an unconventional question in those days, when it was rarely assumed that ordinary people would ever have reason to belly up to a computer keyboard. But Kay already was pondering ideas like people relating to a computer *intimately.* He found himself reading Marshall McLuhan's *Understanding Media* and pondering its seminal koan, "The medium is the message." Then he had his flash of enlightenment, "a shock that reverberates even now," he wrote over twenty years later in *The Art of Human-Computer Interaction:*

> The computer is a medium! I had always thought of it as a tool, perhaps a vehicle—a much weaker conception. What McLuhan was saying is that if the personal

> computer is truly a new medium then the very use of it would actually change the thought patterns of an entire civilization.

Kay was further influenced by a visit to MIT, when he sat in on some of Seymour Papert's experiments with children and computers. Papert, a student of Piaget, had developed LOGO, a computer language specifically designed for children. Eventually, Kay came to the conclusion that the only computers worth designing were those simple enough to be used by children. "If the computer is only a vehicle, perhaps you can wait until high school to give 'driver's ed' on it—but if it's a medium, then it must be extended all the way into the world of the child," he wrote.

To satisfy this rather revolutionary standard, the software in the computer would have to present itself to the user in a totally intuitive manner, enough to entice even the most naive operator into a close collaboration, one in which the user is welcome to manipulate the microworld inside the machine. Whereas previous systems—from the punched cards and batch processing systems of IBM to the dense code words of UNIX—tacked on an interface as an afterthought, Kay understood that future systems would have to be built around a genial software physiognomy. "What is presented to one's senses *is* one's computer," he exclaimed. He and the other PARC people would call it the "user illusion." In other words, the "consensual hallucination" that William Gibson named cyberspace.

To do what Kay wanted, the very language of the

computer would have to be convivial, in a way that no previous computer language was. In fact all previous computer languages, even the supposedly easy ones like BASIC, were a nasty thicket of thorny syntax, geared to the needs of the machine rather than to the habits of human beings. Kay wanted more of the flexibility and creativity of a natural language. He understood how languages shaped the thought processes of the people who used them, and believed that computer languages could do the same. Yet natural languages emerged through an evolutionary process, and thus were somewhat shaped, implicitly yet with unerring sense, according to the needs and desires of its speakers. A computer language had no such advantage: it was thrust upon users in one chunk. Only a very wise designer could come up with something that was intuitively grasped, flexible enough to support complex operations, organized enough to promote new efficiencies, open enough to permit creativity, and fun enough for kids to enjoy.

He'd call it "Smalltalk."

At the very least, it was fun for Kay to design. For a better understanding of intuitive learning, he immersed himself in the works of Piaget and Jerome Bruner, examined the philosophies of Montessori and Suzuki Violin, and even become a disciple of tennis guru Timothy Gallwey, key apostle of the Inner Game of Tennis. (For years Kay's appearances on the lecture circuit featured an amateurish-looking video of elderly women using Gallwey's system to learn an astonishing amount of tennis in an afternoon.) Kay claimed that these lessons could be applied to computer design—it showed that the best interfaces were those free of interference and capable of focusing attention on what was really important—the task at hand.

One thing in particular about the interfaces of that period got the goat of PARC's wizards. This was the practice, almost universal in computerdom at the time, of using "modes." A mode is a current status, a condition. Software designers used them all the time. Most word processors had an *input mode;* when you were in that mode, you could type in new text. But to change the text, you had to exit that mode, using a specific command, and enter *edit mode.* You constantly had to remember which mode you were in at any time, because something you typed while in one mode triggered an entirely different event in another mode.

Compare this anal retentiveness to a piece of paper, the world's greatest interface. No mode whatsoever. You write text, you doodle, you cross things out . . . quite literally a tabula rasa. But just try to get a computer to be as flexible as a piece of paper, and very quickly you understand that thousands of dollars' worth of silicon and design genius gets lapped very quickly by one sheet of twenty-pound bond.

In his own interface design, Kay strived for the clarity and breadth of a piece of paper. He finally cracked the problem by a sleight of hand called overlapping windows. While Engelbart and his Augmentation workers had pioneered the window, the partitions they had in mind each staked out its own portion of the monitor. Not only was it difficult to keep straight which window one was working in, but the windows wound up competing for the extremely limited real estate on the screen. Kay's solution to this was to regard the screen as a desk, and each project, or piece of a project, as paper on the desk. It was the original "desktop metaphor." As if working with real paper, the one you were working on at a given moment was on top of the pile. You could write happily in that window, or draw, or read a letter. Perhaps you could see corners or edges of those windows previously created. To move to the other windows, you used the mouse to move the cursor out of the window and over the representation of one of the windows "underneath." That window would immediately fill out, giving the illusion that it was "on top."

As Kay later wrote, "This interaction was modeless in a special sense of the word. The active window constituted a mode, to be sure—one window might hold a painting kit, another might hold text—but one could get to the next window to do something in without any special termination. This is what modeless came to mean for me—the user could always get to the next thing desired without any backing out."

In other words, as far as the computer was concerned, it was in a mode—but the user felt the freedom of modelessness. Switching windows was natural. If you view things a certain way, we're always switching modes, in everything we do—it's a mode, mode world. But in non-computer life, we don't have to stop and recite a secret word every time we change activities—look down from the television to the newspaper, watch the baseball and swing the bat—we just do it, and thus don't worry about what mode we're in. Kay understood that our computers should allow us that freedom, too. And Smalltalk did.

Later, PARC's scientists developed word processing features that extended these ideas. Instead of having to switch between "input" and "editing" and "deleting" mode while editing, the user would make the text selection first—by pressing a button on the mouse and dragging the cursor over the words in question. (This in itself was a huge improvement over mouseless word processors, which required you painstakingly to mark the cursor positions at the beginning and end of the passage you wanted to cut or move.) The text in question would be "highlighted" in reverse video, like a photo negative. Only then would you inform the computer what you wanted to do with the text—and this would be done not by a command, but by pressing a button on the mouse that triggered a little "pop-up" window to appear near the selected text. This new window was in the form of a menu—a stack of single words that represented commands—like copy, cut, paste, and undo. Moving the mouse over the command and releasing the button caused the action to occur. The sequence of actions is much easier to perform than describe—it takes only a few hours of working with the system to become adept. And the modes were gone. A major barrier between human beings and digitally stored information had been lifted.

But for most people this style of interaction would remain unknown until it became part of Macintosh computing.

Kay's team worked on Smalltalk through most of the 1970s. The original idea was to use it as the operating system of Kay's dream computer, the Dynabook. "A dynamic media for creative thought," Kay called it. "Imagine having your own self-contained knowledge manipulator in a portable package the size and shape of an ordinary notebook. Suppose it had enough power to outrace your senses of sight and hearing, enough capacity to store for later retrieval thousands of page-equivalents of reference materials, poems, letters, recipes, records, drawings, animations, musical scores, waveforms, dynamic simulations, and anything else you would like to remember and change."

But it wasn't too long before Kay ran up against the impossibility of producing a Dynabook until close to the millennium, if then. The state of technology in 1971, as it would be twenty years later, was insufficiently advanced to implement Kay's ideas. Instead of sulking about this unpleasant reality, Kay kept talking about the Dynabook, and word about it spread so widely that it is probably one of the most influential computers in history, though it has never been built. The Dynabook turned out to become less a real object than a *vision* for an object. Everyone in the industry knows what a Dynabook is, and regards it as a sort of technological bull's-eye to aim for. Indeed, Macintosh itself was explicitly designed as something that would evolve into the Dynabook. (Steve Jobs once fed a slogan to his team: *Mac in a Book in five years.* It took eight.)

The Alan Kay style of virtual designing, which he continued long after visualizing the Dynabook, consists of creating imaginative abstractions of what *can* be, going through the motions of gathering a team to build the thing, and discovering important new techniques and innovations in the process. The real product is the body of ideas that circulate from the vision. Kay himself has conceded that technological wizards generally fall into two categories: the Michelangelo types who dream of Sistine Chapels and then actually spend years building them, and the da Vincis, who have a million ideas but seldom finish anything themselves. In this bifurcation, of course, Kay was the ultimate da Vinci.

There *were* Michelangelos at PARC. In a shockingly brief time, they managed to produce a working computer that ran Smalltalk, sort of an interim Dynabook. This was the Alto. Butler Lampson, a Harvard-trained physicist-turned-computer-scientist shared Kay's vision of transforming computers from the exotic to the personal. Though he knew that Kay's design specifications for the Dynabook were entirely too demanding, he also understood that recent advances in technology would enable relatively tiny machines to do heavy-duty computation that, paradoxically, would make them simpler to use. Lampson wanted the Alto to be as powerful as a minicomputer, but less expensive, devoted to a single user, and, above all, radically more intuitive than any machine that preceded it. It would have some of Engelbart's innovations, but a much less imposing learning curve. Like Kay, Lampson got misty-eyed at the possibilities, which included swarms of Joe Sixpacks transmogrified into computer programmers:

> Millions of people will write non-trivial programs, and hundreds of thousands will try to sell them. Of course the market will be much larger and very much more diverse than it is now, just as paper is more widespread and is used in many more ways than are adding machines. Almost everyone who used a pencil will use a computer and although most people will not do any serious programming, almost everyone will be a potential customer for serious programs of some kind.

Within about four months of intense work in late 1972 and early 1973, Lampson, with PARC hardware wizard Chuck Thacker and other engineers, produced prototype Altos. Almost to underline the difference from its straitlaced predecessors, the first image to appear on an Alto screen was Cookie Monster, the voracious biscuit biter from PBS's "Sesame Street." (Alan Kay's group had used the image as a test pattern.) Here was a truly powerful computer, and its first offering was something *personal*—whimsical, even. It was a gauntlet tossed down to the concept that computer time was so precious that not a moment should be wasted. The Alto was different, geared to forgive the meanderings and peccadilloes of an individual user. What excited the PARC-oids was that, unlike the time-sharing machines they were used to, the Alto ran just as fast during the day as it did at night. (Experienced programmers had grown accustomed to late-night computing binges, when fewer people would compete for the cycles of time-sharing computers.)

It's an empty exercise to compare the Alto to other computers using machines with statistics like random-access memory and millions-of-instructions-per-second executed by the processor. The important thing to know is where the power and the memory of the machine was directed: to making a complex graphic display on the screen, and allowing an ultrafast response when the user keyboarded a command, or moved the mouse.

These two abilities are worthy of separate attention.

The Alto graphics were its most striking characteristic. Computer screens in those days, and for many years thereafter, were what was called calligraphic, or character-based. When the user typed a letter on the keyboard, an action much in the spirit of a typewriter occurred—a code was triggered inside the computer memory to generate the letter on the screen, and that information triggered phosphors on the screen to paint that letter. Calligraphic screens generated drawings in the same fashion. This was efficient—it took very little code to display a complicated graphic—but because the electron beams were subject to flicker, the display never looked very good. It looked, well, like a computer, a standard of quality endlessly inferior to that of a printed piece of paper in the cheapest book or newspaper.

The Alto left all that behind, aiming for the intensity of paper and ink. It used *bit-mapping.* Every single pixel on its screen—a spacious landscape about the size of a legal document—was "mapped" to a single bit of the computer's memory. Depending upon whether that bit was flipped on or off, the pixel on the screen would be dark or light.

Bit-mapping required a huge amount of computer memory—the Alto screen consisted of half a million picture elements. That amount required about $7,000 worth of computer memory in every machine. (Thacker understood, however, that this was a temporary expense, and if memory costs fell at the expected rate, that sum would be well under $100 in a decade.) Still, bit-mapping was worth it. It was part and parcel of making computation an experience as intimate as using paper and pencil—and inestimably more powerful. Once you bit-mapped, the computer regarded everything on the screen as a graphic, whether it was an alphanumeric character, a bar chart, or a richly shaded portrait of a human face. For the first time, it was possible—it was assumed—that the display on the screen would be a precise analog of what the eventual output would be when generated on a high-quality printer. It is now common practice to call this feature What You See Is What You Get, voiced as the acronym WYSIWYG and pronounced WHIZ-EE-WIG. (Not particularly mellifluous, but the pronunciation has caught on.)

The other notable feature of the Alto was the full integration of the mouse, which had made its way to PARC along with some refugees from Engelbart's lab. Xerox had done a lot of testing on the mouse. Bill English, one of several SRI workers who left Doug Engelbart to pursue the bit-mapped dream at Xerox PARC, later explained that experiments conducted with people with blood flowing in their veins (as opposed to computer scientists, who bled binary) resolved any doubt remaining about the superiority of the mouse as a pointing device. Fewer than one in ten users had difficulty mastering the rudiments of mouse control. (Why? Basically, using a mouse is like pointing a finger at something.) The befuddled ten percent were presented with a little exercise called Fly. The logic of Fly was simple: there is a fly on the screen that is bothersome and should be exterminated. The mouse controls a fly swatter. Use it to kill flies. Go. "After ten minutes," said Bill English of those who used this remedial training program, "they would be perfectly at home with the mouse."

This easily acquired point-and-click skill could be put to use instantly on an Alto, enabling people to select text and invoke menu commands at will.

With bit-mapping, pointing, and windows, the Alto delivered a welcome alternative to all previous computers. Instead of being challenged by that infuriating command line, all your vital information, your working files, were spread before you like a sylvan vista from a mountaintop cabin. Getting to your work was now as easy as sitting on the porch and pointing . . . *to that tree . . . there!*

As refreshing as a verdant countryside might be to a deskbound knowledge worker, this was obviously not the ideal metaphor for a screen display. The key to an intuitive system in a so-called "graphical user interface" was familiarity. And the way to accomplish this was by a suitable metaphor.

Metaphor, it turns out, is the key to making computers comprehensible. It was not until the late 1970s when two Harvard Business School students named Dan Bricklin and Bob Frankston used a metaphor easily accessible to people who worked with money—accountants, economists, bookkeepers, and anyone who ever wrote a business plan—that personal computers crossed the line from a hobbyist obsession to a compelling tool. The metaphor was that of a spreadsheet—the grid of rows and columns of figures by which one calculated profit and loss. Their electronic spreadsheet was called VisiCalc, and it had many advantages over its paper counterpart, not the least of which was that it liberated users from tediously having to recalculate the entire spreadsheet to reflect changes caused by changing a single number. This freedom allowed people to experiment without penalty, and actually changed the perception of a spreadsheet from a document of hard costs into a modeling tool by which one tested business scenarios. The software metaphor was not only superior to the real thing . . . it *became* the real thing. Now, when people speak of spreadsheets, they do not refer to the green graph paper where spreadsheets used to live—those are useless now. Spreadsheets are tools on a computer.

Some years after the development of Smalltalk, the scientists at PARC came up with a brilliant metaphor for computer screens. Like all great software advances, in retrospect it was obvious: the very desktop to which the knowledge worker was bound. It was a considerable extension of Kay's idea that overlapping windows were like papers on a desk—in the final Xerox implementation, on a computer called the Star, there were icons representing file cabinets, printers, in- and out-boxes, file folders, and of course, paper. Instead of choosing a file from a directory, you would actually *see* the file. Instead of invoking a command to print, you could actually move the cursor to the printer.

The office metaphor was indisputably one of Xerox PARC's greatest leaps. (Though Alan Kay's original desktop metaphor involved only the concept of using overlapping windows to represent papers on a desk, the term "desktop metaphor" quickly came to stand for this broader concept including tools throughout the office.) Like many other brilliant ideas, once introduced it is unimaginable to conceive of working without it. Smalltalk could potentially offer an infinite number of scenarios, each of them a new world with windows opening into the informationscape. What better way to emulate the sort of work that most of us do with computers—deskwork—than by making a virtual desk, with virtual drawers, virtual file folders, virtual paper?

Though the Alto was one of the truly revolutionary information processing products of the century, the lords of Xerox were never quite sure what to do with it. As documented in *Fumbling the Future,* a book about Xerox by Douglas K. Smith and Robert C. Alexander, there were plenty of (lame) reasons for this failure, from corporate politics, to timidity, to outright misunderstanding the marketplace. Certainly every time the Alto was exposed to the world outside the sleek monastery near Route 280, the response was terrific. Xerox placed a few Altos in the Carter White House, but did not publicize the enthusiastic reaction. And when Xerox conducted an experiment whereby a group of secretaries used Altos equipped with a word processor named Gypsy that exploited the computer's friendly virtues, the result was astonishing. Within a few hours they were working productively on it. Still, Xerox surrendered the word-processing market without firing a shot, allowing companies like Wang to dominate with systems that were clunky, overpriced, and hideously arcane.

Xerox eventually did make an attempt to sell PARC's concepts in a real product, the aforementioned Star. Adopting the paradigms of Smalltalk and the Alto, and drawing from the experience of users exposed to the thousand or so prototypes of that machine, the Star's designers considered their creation a potential world-changer. "These paradigms *change the very way you think,*" they wrote in an article in *Byte* magazine, "They lead to new habits and models of behavior that are more powerful and productive. They can lead to a *human-machine synergism.*" (The emphasis is theirs, apparently a case where the writers yielded to the ease with which one could generate italics with the Star.)

In retrospect these wild claims are well in line with reality. Yet it was not the Star that presented these paradigms to the masses; it was the Macintosh and, later, its imitators. Xerox blew it again. It had no idea how to sell the Star, which at $18,000 for a basic model and much more for a fully loaded setup, was an awkward buy for a corporate manager. As a Xerox executive put it succinctly to the authors of *Fumbling the Future:* "[The Star] was a technological *tour de force*—but it was too expensive, no one understood it, and no one wanted it."

Though the PARC researchers were understandably frustrated by Xerox's inability to introduce their advances into the marketplace, they weren't exactly despondent. By and large, they viewed themselves as pure researchers. A prototype, a paper, an article in *Scientific American* . . . those were their products. They had little faith that the hoi polloi would very soon be using computers. In contrast, a small group called the Homebrew Computer Club met twice a month at the Stanford Linear Accelerator Center—not more than an Aerobie toss from PARC itself—discussing and implementing the beginnings of a personal computer industry. One of the Homebrew members was Stephen Wozniak; he designed the first Apple computer to impress his fellow clubsters. When PARC scientist Larry Tesler attended a Homebrew meeting in 1976, his disdain was indicative of his peers' attitude: "I watched guys carrying around boxes of wires and showing programs that generated flashing lights. My neighbor said, 'This is the future!' . . . I told him, 'Forget it.' "

Compared to the sophisticated digital wonders at Coyote Hill Road, the Homebrew hackers were playing with arrowheads and sticks. What the PARC people did not understand, however, was that people accustomed to performing tasks with their bare hands would consider even primitive tools like arrowheads and sticks miraculous.

Tesler would soon change his mind both about personal computers, and his priorities: he came to realize that actually shipping a product was as important as conceiving it, and thus the concept of making idea reality moved up several notches in his cerebral stack. But others at PARC never really came around. I find the purest expression of PARC's elitism in this comment by PARC hero Butler Lampson, when interviewed by Susan Lammers for her *Programmers at Work:*

> Were we aware of the outside world? Yes, we knew that it existed. Did we understand that whole situation entirely? Probably not. Were we surprised when Xerox was unable to sell Stars? No, not really . . .
>
> The purpose of PARC was to learn. You owe something to the company that's paying you to learn, and we felt we should do what we could, within reasonable bounds, to benefit Xerox. But it wasn't critical that Xerox develop those ideas. Their failure was not really surprising . . . Some things went wrong in marketing . . . That sort of thing is annoying, but the main product of a research laboratory is ideas.

The scientists at Xerox PARC proceeded as if they were working with ARPA grants, not corporate funds. They were satisfied to learn, to come up with new ideas, and to nudge the field forward. And if, as Lampson explained, it could be done "within reasonable bounds" they would actually do a little something for the company paying their salaries.

This was stark contrast to the scrappy engineers at Apple Computer, whose job it was to create new products that would both change the world and bring in some cash. The ethic was more practical: if it didn't hit the streets, it wasn't worth doing. Ideas were useless if they didn't get out there. For the Macintosh, the most important design consideration was getting it on people's desks.

As for new ideas, yes, the Apple folks cooked up a thick stew of them. But the best ones were borrowed. These were, of course, the ideas concocted and nurtured by the computer science Illuminati at the Xerox Palo Alto Research Center.

4

You can have your Lufthansa Heist, your Great Train Robbery, your Crown Heights Caper, and your Brinks Job. For my money, the slickest trick of all was Apple's daylight raid on the Xerox Palo Alto Research Center.

It was December of 1979 when the princes of PARC received a group of visitors working on a new computer for Apple, to be called Lisa. The unusual excursion was the culmination of some high-level negotiations between Steve Jobs and the people at Xerox Development Corporation, a branch of the copying giant specializing in venture capital. Apple was in the midst of soliciting new investments to sustain its meteoric growth, and Xerox's investment arm was interested. Here was the eventual deal: in exchange for allowing the Development Corporation to buy 100,000 Apple shares for $1 million, Xerox would host a contingent of Apple engineers for a peek at PARC's wonders. The number crunchers at Xerox considered this a fairly innocuous concession—they were getting a tangible stock deal in exchange for allowing Apple a brief exposure to technology that in their minds belonged more to science fiction than to future revenues.

Even the PARC people didn't think much about the impending visit. Demos to outsiders, if not an everyday occurrence, were not unusual. One person designated to show Smalltalk and the Alto to the outsiders was Larry Tesler, chosen largely because he was one of the few engineers who thought that personal computers were anything other than a joke. (A grudging change from his first visit to the Homebrew club three years previous.) Tesler had actually purchased a few early PCs. Still, at the time he regarded those working in the field as belonging to a lower class of technologists than he and his peers. "These were a bunch of hackers, and they really didn't understand computer science," he told one journalist. "They really wouldn't understand what we were doing and just see pretty dancing things on the screen."

The eight representatives of Apple were ushered into a demo room. There were Steve Jobs, Bill Atkinson, Apple's president Mike Scott, and various executives and engineers on the Lisa project. Tesler flicked on the Alto. The paradigm of augmentation came alive—windows, bit-mapping, icons, pop-up menus, mouse. The Apple people might not have been trained as computer scientists, but they had an instinctual sense of what was good technology. This was very good technology. They wanted to know everything about it, how it was done, what were the limits. Science fiction writer William Gibson's oft-noted aphorism applies here, sort of: "The Street finds its own uses for things—uses the manufacturers never imagined." Already, the junkyard mentality of Apple engineers was hard at work, imagining what could be done with the concepts before them.

Often at PARC demos, newcomers failed to comprehend the marvels before them. They would rave for twenty minutes about the mouse, and totally ignore the significance of bit-mapping. The scariest response came once when, after a virtuoso demo into the frontiers of computer science, the lucky recipient commented, "You really get good reception here." But the Apple people were incredibly sharp. Tesler would later admit that in the seven years he had been at PARC, nobody, whether a casual visitor, Xerox executive, or computer scientist, has queried him with such insight. (He later learned that they'd studied the Smalltalk literature before the visit.) "It was almost like talking to someone in the Group," he later told me. "But better, because they wanted to get it out into the world." Tesler was so impressed, in fact, that a few months later, despairing of Xerox ever getting its act together, he left PARC to seek employment in the personal computer world. At Apple, of course.

Tesler recalled Bill Atkinson sitting with his face almost pressed against the screen. Meanwhile, Steve Jobs could hardly contain himself with excitement. After he watched Tesler manipulate the screen display, open windows, click on icons (the Apple people were not permitted to handle the goods), he nearly exploded. "Why aren't you doing anything with this?" he bellowed. "This is the greatest thing! This is revolutionary!"

Steve Jobs, Bill Atkinson, and the others walked out

that day with something much more valuable than diamonds, treasury bills, or even gold bullion. A paradigm. By the time Xerox noticed they had the idea, it was already much too late. Apple had gone off to start the revolution without Xerox.

The Engelbart-PARC paradigm was the bedrock from which Apple was to construct the future, first with the Lisa, then the Macintosh. The payout so far? As Carl Sagan might say, billions and billions. Dollars aside, the true significance of that day was that for the first time, the wonders of augmentation, of informationscape, of working in a world of metaphor, were demonstrated to people *who would do something about it.* The drive back to Cupertino was only fifteen minutes or so but before it was over the Apple people were already discussing how to change the world with the ideas to which they had just been exposed.

How long do you think it will take? Steve Jobs asked Bill Atkinson.

"About six months," Atkinson replied. He was only off by three years or so.

At the time of the Xerox visit, Apple Computer, Inc., had been incorporated for less than three years, but was already the world's leading personal computer company. But it was much more than that. From its start, Apple was never a mere corporation—it was a symbol. Depending upon your point of view, it could symbolize the information explosion, the entrepreneurial revolution, or the maturation of post-sixties baby boomers.

Its roots were legendary. College dropout and former engineer Steve Jobs had a savvy knowledge of technology and a bent toward promotion. And Steve Wozniak, aka Woz—Jobs's friend from Homestead High School in Cupertino, California—had a computer. Jobs, then twenty-two years old, not only convinced Wozniak to leave his job at Hewlett-Packard to form a company with him, but the pair managed to recruit some savvy venture capital, as well as the support of the key public relations firm in Silicon Valley.

Steve Jobs's garage was the company headquarters in 1976 when the two Steves, aided by a couple of high school kids named Chris Espinosa and Randy Wigginton (both of whom would work on Macintosh), were preparing the Apple II computer for its debut as a product early in 1977. (The Apple I had been a single circuit board, sort of a prototype for the real thing.) The garage was fun—a few hackers making an extremely neat machine. Wireheads and hackers were uniformly impressed by Wozniak's virtuoso design. They regarded its motherboard, the main circuit board, as a beautiful work of art.

But the Apple II had broader appeal: it was the first personal computer capable of being appreciated by a wider audience than hobbyists. It could output colorful graphics on the screen, had a solid version of the BASIC computer language built in, and was housed in a friendly plastic casing. Those who spent time with the machine also came to see a special magic in it; somehow the insouciance and verve of its designer came through. Most of all, it was a real computer, designed for use by an individual, and priced not at $100,000 (as most people assumed computers cost) but under $1,500. It was the main force in compelling people to consider the possibility of having something as exotic as a computer on their own desktops and thus, it was destined to become a landmark in the industry, selling millions, and making the Apple name virtually synonymous with personal computers (until IBM entered the field). It was also to make both Steves very rich and famous.

"It was like inventing a new form of transportation," recalled Chris Espinosa. "A computer was something that had been heretofore seen, but only in a larger, more expensive, more cumbersome form that was reserved for the elite. [This was like] raining gold upon peasants. There was a sense that we were on the cutting edge of something new and important and that we could change the world with what we were doing. Very few of us realized that changing the world had financial implications."

As a consequence of the company's success, Apple very quickly had to shift from a garage mentality to the mindset of a budding corporation—one valued, at the time of the PARC visit, at over a billion dollars. It filled several low-slung office buildings in Cupertino, and had hundreds of employees. Though the Apple II was wonderful for its time, Apple's leaders realized that the company needed new products to remain competitive. They began work on the Apple III, a machine roughly as powerful as IBM's personal computer would be.

But Steve Jobs had an idea for something even more special—Lisa, a computer that would leapfrog Apple's technology, surpassing not only the Apple II, but Apple

III as well. This jump would also vault Apple a generation or so past anything that its competitors were preparing. Begun when Steve Wozniak, at Steve Jobs's request, sketched its architecture on a napkin, Lisa had, in less than a year, evolved to a computer based on the powerful Motorola 68000 microprocessor chip, and was engineered to handle more complicated applications, even running several at the same time, a trick called "multitasking."

Named after an Apple engineer's daughter (and allegedly an additional tribute to Jobs's own daughter), Lisa was also the first Apple computer specifically directed at office professionals—from white-collar workers to captains of industry. If you picture the Apple II as a Volkswagen bug, whose speedometer stopped at 90, the Lisa was a German sports car, with a needle that topped out at 180.

To produce Lisa, Apple had hired away engineering and management talent from a serious computer company, Hewlett-Packard. At the time of the Xerox visit, Apple had a working prototype of a machine that reeked of serious intent. The model they kept in mind was the Hewlett-Packard 3000, a minicomputer that had never cracked a smile in its life. Yet from its start Lisa had been inherently interesting—the concept had always been a thorough integration of text and graphics. Lisa's most promising feature thus far was the work of Bill Atkinson—fast software routines for bit-mapped graphic output.

Atkinson had also, over the objection of some of the

hardware engineers on the project, successfully lobbied for Lisa to operate on a "paper" paradigm. That meant that the background color of the screen would be white, not the null void of almost every other computer (except the Alto). The hardware people hated this because a white screen was prone to video flicker. To avoid the problem, the computer would have to refresh the phosphors on the screen much more often. This process would require Lisa to become more powerful, and therefore more expensive.

But the result would be worth the cost. Lisa would presumably encourage office workers to produce documents blending text and graphics. If the screen background were black, the graphics would resemble photo negatives; people would be unable to visualize what they really looked like until the document was printed. Also, people were comfortable putting dark marks on white paper, and a computer that provided them with that comfort would be a welcome departure from the current, vaguely hostile dark-background computer monitors.

In the excitement generated after the PARC visit, Jobs and one of his lieutenants, Trip Hawkins (who later would found Electronic Arts, a software company that devised some extremely slick computer games), wrote up a new plan for the project, focused on implementing the key Engelbart-PARC innovations—windows, icons, mouse. Though the computer would still be marketed to office professionals, this represented a different philosophical underpinning. Lisa was now destined to vivify what personal computers *could* be. By elaborating on

Xerox's paradigm, Apple would have a computer that instantly rendered every single one of Apple's competitors an antique.

This proposed shift in course was not a given. Even though Jobs, as cofounder and chief stockholder of Apple, had considerable power, it was understood that his youth and impetuousness relegated him to a subordinate role to Apple's chairman Mike Markkula and president Mike Scott. To the latter, the opinions of the former Hewlett-Packard engineers on the Lisa team were regarded at least as highly as those of Steve Jobs. And the H-P people thought that, as impressive as the Alto was, it was rather presumptuous, if not downright flaky, for Apple to attempt to duplicate the feat in a low-cost personal computer.

Presumption, however, was Jobs's forte. He and those sympathetic to his vision lobbied furiously. With Xerox's permission, they commissioned their own mouse from Jack Hawley, a Berkeleyite who had left PARC to open a business called Mouse House ("Purveyors of Fine Digital Mice to an Exclusive Clientele since 1975"). They dubbed it the "clandestine mouse." Bill Atkinson quickly hacked a driver program that allowed the mouse to move a cursor on the computer screen. Jobs and Hawkins proceeded to dazzle skeptics with the power of the pointing device. The mouse triumphed.

A torch had been passed. The nexus of twenty years and millions of dollars of government and top-level corporate research was now in the hands of a company that only a few years before had operated out of a garage.

Apple's task was to take this technology out of the lab and into general circulation.

Bill Atkinson quickly learned how difficult this would be. His job on the project was to generate the routines that would control the display—the software equivalent of heavy lifting. (An intense, bearded engineer named Rich Page was his hardware counterpart.) At twenty-eight, Atkinson had established himself as one of the true wizards in the company. Before coming to Apple, he had been a graduate student combining computer science and neurobiochemistry at the University of Washington, gaining distinction by concocting a graphics program that interpreted CAT scans of the human brain. The stunning visuals produced by Atkinson's work allowed people to see the brain from previously unimagined vistas. Atkinson had experienced a conversion experience when he came across an Apple II in 1977. He easily saw past its limitations (it was much less powerful than the machines he worked with at school), instead appreciating the virtuosity of Steve Wozniak's digital design. He went to work for Apple in 1978—employee number 51—writing applications that would help sell the Apple II.

But with Lisa, Atkinson faced his biggest challenge. His ninety-minute exposure to Smalltalk had been somewhat deceiving. While the computer in some ways seemed to have completely solved the challenge of allowing an unschooled worker easy access to the information inside the computer—the furniture of cyberspace—when Atkinson sat down and tried to duplicate the task, he realized that there were gaps as yet unfilled. Specifically, there was the question of what was known as clipping. It had to do with the phenomenon of overlapping windows. When you opened a window, resized a window, or moved one, what happened underneath? Did the computer have to perform all the work of drawing the windows you did not see, or was there a way that it could save energy by drawing only the part of the display the viewer saw at a given moment?

During the PARC visit, Atkinson was impressed that Smalltalk somehow "knew" how to show only the visually relevant information at any given millisecond. The irony is that Atkinson was mistaken—the Alto used a much less elegant, and slower, method than clipping. But buoyed by what he thought was Smalltalk's existence proof of clipping, Atkinson kept hammering at a solution. He wanted a system for Lisa whereby these hidden "regions" would redraw so quickly that the user would have the illusion that everything was really on the screen at once.

Atkinson worked at the problem for months—not only in long hours at a desk, but literally in his dreams. Upon arising he would record his somnambulant labors in a notebook. Eventually wave after wave of Atkinson's brainpower eroded the problem. He had set out to reinvent the wheel; actually he wound up inventing it. His solution dealt with a sophisticated use of algebra to calculate which "regions" of the window had to be drawn and remembered.

For a time, Atkinson was the only person in the world who understood the voodoo by which these regions could so quickly display overlapping windows.

This lack of redundancy was almost disastrous. On his way to Apple one morning, Atkinson failed to notice that the tractor-trailer ahead of him was parked. He drove his RX-7 straight *underneath* it, shearing off the top of the sports car, which wound up on the other side of the huge truck. The subsequent call to emergency services reported a decapitation. Fortunately, Atkinson's head, at the time the sole repository of the secret for representing cyberspace on a computer screen, remained in place, though for a brief period, rather dysfunctional. He awoke in a hospital room. Steve Jobs was staring at him with concern. "Don't worry, Steve," he said. "I still know how to do regions."

Atkinson's breakthrough was the bedrock of a set of graphic routines he called LisaGraf, later renamed to the more generic QuickDraw. These were the heart of the Lisa display, and eventually the Macintosh. "QuickDraw made the Macintosh graphic friendly," Atkinson told me some years later. "That's what made it an environment fertile enough to do PageMaker."

In mid-1980, Larry Tesler, late of PARC, joined the Lisa project, and eventually was joined by other Xerox refugees, including Owen Densmore and Steve Capps. The computer scientists made for an interesting contrast with the professional engineers from H-P, not to mention the personal computer hackers who embodied the improvisational Apple style. The Lisa project became characterized by discussion, a carryover from the beanbag-chair colloquies at PARC.

Steve Jobs was no longer with the project. Apple had

reorganized and former H-P manager John Couch had been named the head of the Professional Office Systems Division from which Lisa was developed. The underlying motive for this switch was getting Jobs out of the building—his impulsive management style and overbearing ego were driving people crazy. Also, the company's top executives, cautious in their first assault on the corporate world, were concerned that Jobs, at twenty-five, did not have the experience to sell into that cloistered, conservative market. A few months later, Jobs's disappointment was slightly mitigated by his elevation to Apple's chairman of the board.

By the time Tesler came on board, Lisa's hardware design was completed, or, in geek terminology, "frozen." Lisa now presented a pleasant if beetle-browed visage, with its twelve-inch screen situated to the left of two slots for floppy disks that would run on the ill-conceived "Twiggy" disk drives. (The Twiggies were unreliable artifacts of Apple's hubristic desire to do even needlessly difficult things in-house, instead of relying on technology perfected by outside specialists.) A detached keyboard, in a deep beige hard-plastic casing, rested under a beveled overhang underneath the screen.

But the software, particularly the interface, was up for grabs. The design process was evolutionary. Every aspect of the operating system exposed to the user was subject to a series of incremental improvements; these would be punctuated by clever, sometimes brilliant, mutational innovations. Driving the process was the ongoing debate over what might be the very best face Lisa could put forth. Even the slightest interface aspect could trigger a heated debate, with adherents of opposing solutions arguing with near-Jesuitical intensity. With good reason—when an interface is exposed to millions of users, even the most minor inconsistency can be amplified to a consistently infuriating annoyance.

To this day, the bone that sticks most deeply in the craw of Apple and Macintosh designers is the charge that all their interface work simply consisted of making, no pun intended, a Xerox copy of the work they saw at PARC. Indeed, the same issue was raised in a lawsuit filed by Apple against the software developer Microsoft, when Apple complained that Windows, Microsoft's own visual interface, owed a suspicious debt to Lisa and Macintosh. According to *Gates,* the biography of Microsoft's chairman, when Steve Jobs first accused Bill Gates of stealing Apple's ideas, he shot back, "No, Steve, I think it's more like we both have this rich neighbor named Xerox, and you broke in to steal the TV set, found I'd been there first, and said, 'No fair, I wanted to steal the TV set!' "

A look at the evolution of the Lisa interface, however, shows that much more was involved than lifting a Trinitron from Xerox's living room. With the discipline of the marketplace looming over them, Lisa's engineers realized that PARC's ideas had to be stripped down and rebuilt to more demanding specifications. By the time Tesler arrived in mid-1980, Apple already had clarified some of PARC's ideas, making them friendlier to novice users. One of the primary differences was the implementation of something called direct manipulation—the ability to reach into cyberspace and get things done without any mediation. In the PARC world, things mostly got done by moving the cursor over selections on pop-up menus. With Lisa, however, you could manipulate almost anything on the screen, often without reverting to the middleman of menus.

"Xerox never had the concept of direct manipulation," Tesler later explained. "Even with the Star you couldn't drag a window around. You couldn't drag an icon around. To resize a window you had to use a menu. Atkinson had the idea of dragging these around." As a former member of Alan Kay's Learning Research Group, Tesler understood the irony—striving to make computers accessible, Kay had been designing for children, and had steered the nature of computer interaction away from obscurity, on the golden road toward intuitiveness. But he didn't go far enough. It took the street hackers of Apple to go the final mile. "Alan Kay thought Smalltalk was for kids, and it was simple—a fourteen-year-old could do it. But Bill wanted it *simpler,*" Tesler later recalled.

Tesler's comment reveals the extent to which Bill Atkinson influenced the design of Lisa's human interface. Because Atkinson wrote the software that controlled the objects on the screen—the visual part of the interface—he had a strong voice in debates. When Tesler arrived, he became chairman of the Lisa User Interface Council, consisting of himself, Atkinson, and the most receptive marketing representative they could find.

A pattern emerged. An engineer, usually Atkinson

or one of his cronies, would propose an unusual innovation. The marketers, fearing yet another time-consuming rewrite of the software, would object. They would be joined by the H-P contingent, a cautious lot sensitive to all sorts of blasphemies against previously held wisdom. At the point of impasse Tesler would turn to Atkinson and ask how long it might take to implement the feature, quick and dirty. Tomorrow! Bill would say. The next day, after an all-night hack, Atkinson would have a prototype. Then they'd test it.

User testing was a Larry Tesler's fetish. Since this was supposed to be the first computer that virgin users could begin using as soon as it landed on their desks, the people testing could not yet have been initiated into the mysteries of digital culture. Fortunately, Apple in those days was hiring boxcars full of new employees, many of them as secretaries, janitors, marketers, and accountants. They'd never touched a computer. Tesler had a connection at the new-employee orientation sessions held on Mondays who recruited newcomers for user testing sessions at Lisa's Bandley 2 headquarters, "before they got contaminated" with exposure to other computers, Tesler later noted. He'd sit a virgin before a Lisa and conduct controlled experiments on isolated features. First do it this way . . . now that way. After four or five testers had slithered through this interface maze, the correct solution would usually emerge.

This give-and-take system yielded some marvelous digital artifacts that distinguished Lisa, and later found their way to Macintosh. Take the menu bar, that row of words that rests on the white space at the top of every Macintosh application. If you move the cursor over one of the words, like FILE, EDIT, FONT, or SPECIAL, you get Apple's successor to Xerox PARC's pop-up menu—the pull-down menu. This drops like a window shade, with a list of words representing a command that will be enacted when the cursor finds its way over the proper word.

As it turns out, Bill Atkinson did not originally plan for a menu bar on top of the screen. It sort of migrated upward, like cream rising to the top. Atkinson wanted the commands to be geographically predictable, the same place in every application. For that reason, he rejected pop-up windows. Then he got the idea of a menu bar—a constant presence from which one could evoke a menu of commands by pointing and clicking. At first he put the bar at the bottom of the active window, then after some testing moved it to the top. But putting the menu bar on the top of a window presented a problem: when the user shrank the size of the window, you couldn't get the headings to fit. (Microsoft's Windows uses this approach, and suffers by it.) So finally, at Tesler's urging, he moved it to the top of the screen. Users quickly learned where the commands were under the individual headings and Atkinson was impressed that they could implicitly visualize where a command might "live" on the screen, moving the mouse over the heading, dropping down the menu and going right to the location in one fell swoop. It was a case of totally internalizing the illusion of geography in cyberspace—you would "go" to a menu choice that didn't really exist until you created it.

Then there was the mouse button war. To outsiders the issue of how many buttons on a mouse is as arcane as how many angels can be jammed on the head of a pin. But debates about mouse buttons are impassioned and urgent. As Bill Gates once told me, "The number of buttons on a mouse is one of the most controversial issues in the industry. People get religious."

In 1971, when Bill English and a couple of other SRI workers left Doug Engelbart to pursue the bit-mapped dream at Xerox PARC, the mouse retained its three buttons: red, yellow, and blue. But in preparation for the Star, Xerox tested a number of mouse variations on new users, and found two buttons to be a felicitous number. One button was used for selecting things, the other button to extend the selections.

As had Xerox, Apple tested many variations of the mouse, experimenting not only with buttons, but the mouse's size and shape. Ergonomic comfort was important; in five years of use, the modest increments of mouse travel on a desktop or pad could total up to a journey between twenty-five and thirty miles. Industrial designer Jerry Manock once estimated that they tried 150 models of mice, broken up in "wine-tasting sessions" of around twenty-five each. One variation was shaped like a golf ball. Apple finally decided that smaller was better, and shagged the so-called Arnold Palmer model. (While the initial version of the Macintosh mouse was indeed tiny compared to the brickish Engelbart original, it was still comparatively boxy, and Apple standardized a sleeker version some years later.)

The results were more interesting in the button competition. When Apple tested a two-button mouse, Tesler recalled, "people made a lot of mistakes. With two buttons, they'd constantly be turning their heads from the screen and looking at the mouse. I'd say, 'Hit the button,' and they'd say 'Which button? Which button?' When we did experiments with identical everything, except the number of buttons on the mouse, the people who used a one-button mouse said it was easier to pick up. I realized that when you used the mouse, they were pointing. Pointing, and tapping the button with the finger they were pointing with. There was no mental model for pointing with more than one finger. So we got rid of the second button. I wrote a memo called 'One-Button Mouse.' "

Unlike some of his former colleagues, for Tesler the conversion to a single button did not engender a spiritual crisis. Apple's interest in Lisa, he would later explain, was "getting systems to laypeople, not computer experts." His apostasy was repulsive to the Xerox people. The one-button mouse "was a mistake," Bill English said. "Why do you want to limit yourself like that?"

But Apple's mouse really wasn't limited. Instead of multiple buttons, the Lisa mouse—and the Macintosh mouse that followed—let you use the single button in different ways—by the number of times you clicked the button. There were single-clicks and double-clicks. Like a coded shorthand—one if by land and two if by sea—the Macintosh listened to see if the user clicked once or twice on the mouse. As with many of the skills required to use the Macintosh, this may sound complicated but is actually rather intuitive.

The word processing program at Xerox had used double-clicks to select words, but the Lisa group used that function for other things as well, particularly for launching applications, and it was further refined on Macintosh. One learns rather quickly which commands are double-clickers and which are single-clickers. Selecting a file is a single-click. Launching the file requires a double-click. In text editing, moving the cursor is a single-click. But if you place the cursor on a word, double-clicking the mouse highlights that word. A few years after the Macintosh's introduction, a group of third-party programmers designing a hot new word processor decided to implement *triple*-clicking—click three times, fast, and an entire paragraph would be selected. (This evolutionary wrinkle caught on, and now is a semistandard used by several word processing programs.)

Do people sometimes fail to hit the button quickly enough for the second click? They sure do. Do trigger-happy mousers sometimes do an index-finger stutter and double-click when they only meant to click once? Yes. Double-clicking isn't a perfect tradeoff for the simplicity of a one-button mouse, and the mouse button debate still rages.

On the other hand, I always find it extremely comforting to actually *see* the result of a double-click launch. That was one of the great things about both the Lisa and the Macintosh. Apple makes it into a little entertainment event, complete with animation, when the window zooms out from the icon like the beginning of a Looney Tunes episode. That process—of moving the pointer with the mouse over an icon representing an application, double-clicking, and watching a work window open up—was only one of many original innovations that the Lisa interface team added to the bedrock of the Smalltalk paradigm.

The wildest change did not come until 1982, when Lisa was almost ready to have its code frozen at last. It concerned the Filer, the program that would greet every user as soon as he or she flicked on the machine. From Filer, you would launch programs, copy files, name them, and perform other housekeeping tasks. Yet as the shipping date approached, many of the designers were unhappy with the Filer. Every time a user named a file, saved a file, or sought to open a previous file, Lisa would begin a sort of visual interrogation, requiring the user to click off the answer to three questions. This came to be known derisively as the "20 Questions interface," and though it was powerful in conducting searches—it actually was the front end of a minidatabase—some of the engineers considered it unwieldy. (The more acerbic critics called it the "*Hundred*-Questions interface.") The final straw came when Dan Smith, the main implementer of 20 Questions, tested it on his wife—and *she* couldn't get it. He went to Atkinson and, with engineer Frank Ludolph, they began collaborating on something radically different. Since any change that late in the game could be costly—too costly for the marketers to consider without Valium—they proceeded with stealth. "Don't tell me what you're working on, but good luck," said Wayne Rosing, Lisa's director of engineering. With that vague sanction, the three engineers vanished, working mostly at Atkinson's house, until they reemerged with the Desktop Manager, a totally new interface, which possessed a degree of direct manipulation that surpassed anything ever seen at Xerox, ever seen *anywhere.*

Icons took new importance on Desktop Manager. As the name implies, the desktop metaphor had fully flowered. Just as on the real-life desk stretching before the user, the virtual Lisa desk could be cluttered with documents, folders, and tools. (If you wanted to clean it up there was a command that rearranged things in orderly rows.) The tiny hieroglyphs represented either a document, a folder, an application, even a floppy disk. When you wanted to begin a new document you would use the mouse to "reach" over to a stack of blank papers, double-click, and a window would open representing the page before you. Even the "file delete" function was integrated into the make-believe office. On the lower right corner of Desktop Manager was a picture of a trash can. Dragging a file over it would, in effect, delete the file. But not immediately. If you moved the cursor over the trash can and double-clicked, a window would open, and all the icons of the files you had just discarded would appear. You could drag any or all of them to safety. Only when new items were discarded would the trash can be figuratively taken to the dump.

The ability to open a file simply by moving a mouse—sticking one's hand in cyberspace—and selecting an icon was amazing. Compare this to the task of opening a file in the oppressive operating systems that dominated the industry until Macintosh—CP/M and DOS. These command-line systems forced you to type the entire name of the file you wanted to open. The number of characters in the file was severely limited, and often the names you assigned to the files were so cryptic that you would spend a few minutes pondering what it might contain (Julia2? *Who's Julia?* Is there a Julia1?) and wouldn't solve the issue until you opened it up and saw its contents. If you mistyped a single character of the file (as I often did), the system would assume you were not opening a previous file, but beginning a brand-new file with the typo as its title. And you would sit there, frozen, for twenty seconds that seemed eternal, as the computer went through its paces so you could then delete that unwanted file and try once more to get the file you desired. Have you ever gotten on an elevator on the third floor, intending to visit someone on the tenth floor, and then, by instinct, pushed the Lobby button? And waited in rage and self-recrimination as the elevator began its irrevocable descent? Manipulating files on personal computers was worse, much worse. You went the wrong direction five times a day!

But with Lisa, all you had to do was scan the monitor for your document, and *get it.* The change was so significant that it actually changed the way people thought about their documents. No longer was there a barrier between them and their information—no stern homunculus of a code word was required to get your document. The document was right there! You could see it. The old way of thinking—that computers somehow stored vague chunks of information and by operating them you were some sort of accountant or data processor—no longer applied.

When this power was extended on the Macintosh, the stage was set for a generalized rethinking of our relationship to information. The use of metaphor was so effective that, at some point, it was no longer clear where metaphor ended and reality began. As linguists Lakoff and Johnson explained in *Metaphors We Live By,* reality itself can be shaped by metaphor. For instance, our acceptance of the metaphor "Argument is war" and attendant expressions like *attack a position, gain ground, indefensible,* and so on, can actually shape the way we conduct our arguments—like conflicts, with winners and losers.

But Lisa, and later Macintosh, used the power of the digital computer to exploit metaphor in an equally powerful manner. "Most Macintosh users," observed Thomas D. Erickson of Apple's Advanced Technology Group, "believe that when they move a document icon from one folder to another, they are moving the document itself." He goes on to explain that what "really" is happening is that some bits are being changed inside the computer so that "a pointer to the file" is being moved. But I don't see why this negates the users' belief that they are really moving a document. Just because the motion occurs in cyberspace does not mean it is not "real." What we are concerned with when we move a physical file from say, a typewriter to a manila folder in a file cabinet, is not the rag content of the paper, the stiffness of the folder, or even that paper is involved at all. We do it to move information from one location to another, so we can retrieve it later. That's the essence of the action. For all practical purposes, when Macintosh users drag a document icon from one folder to another, they *are* moving a document. Just as when they drag an icon representing a file over a trash can, they are emphatically throwing out the file. Not just "pretend" throwing it out, or virtually throwing it out. It's gone! Just as lost as if you had tossed it into the wastebasket.

Interestingly, when the Desktop Manager was submitted to user testing, the human guinea pigs who used both versions found neither of them easier to learn than the other. The error rates were about the same with both versions, too. But the testers all agreed that of the two, the later version was by far the more enjoyable to use. Tesler would hear comments like, "I like the little pictures." Atkinson's vindicated boss presented the Desktop design triumvirate T-shirts that read, "Rosing's Rascals." The Lisa interface was complete.

All in all, Lisa was a spectacular achievement. Apple had spent $50 million in development, but as in a good Steven Spielberg flick, you could see all those dollars on the screen. Still, by the time Lisa was ready for its public rollout, even the designers knew that, at the very least, Apple had a tough sell ahead. Lisa simply cost too much. The designers had once hoped to sell it at $2,000. But with every improvement the price rose. The black-on-white display. Bit-mapping. Mouse and windowing. Multitasking. These all required expensive hardware, and ate up huge gobs of memory. Lisa shipped with over 1000K bytes of memory, known as a megabyte, and that was barely enough. Lisa also proved unable to run adequately on floppy disk drives alone, and had to be shipped with a hard disk drive, an expensive rarity for a personal computer of its time. The final cost? Over $12,000. And Lisa ran painfully slowly, not as slowly as the Xerox Star, perhaps, but still at a pace more glacial than frisky. It was not that noticeable in demos, but was an infuriating reality to those who used it.

If Lisa had cost half as much, and was several times as fast, perhaps Apple would have been able to market it successfully. And then again, perhaps not. Apple had never really learned to sell computers to Lisa's target, Fortune 500 corporations. And it did not start with Lisa.

In January 1983 Apple announced Lisa. (It did not actually ship until early spring.) Steve Jobs and the newly hired president of Apple, John Sculley, traveled around the country, performing demos for the press and potential corporate customers. The media greeted Lisa warmly, noting that its steep price still represented a dramatic drop compared to what it might have cost previously for such an advanced example of technology. They hailed Jobs, and wrote glowingly as well about Lisa's project manager, John Couch. (They wrote almost nothing about key designers Bill Atkinson or Rich Page.) Everyone agreed that Lisa represented the future.

But not the future of Apple. That would belong to a smaller project, working to create something referred to as "Lisa's little brother": Macintosh.

5

In an industry where the term Renaissance Man is tossed around as casually as a Nerf ball, the man who began the Macintosh project was the real thing. Jef Raskin had degrees in computer science and philosophy. He was a painter whose work had been hung at the Museum of Modern Art. He was a musician, a former conductor of the San Francisco Chamber Opera Company. For five years he had been a professor of visual arts at the University of California at San Diego. He celebrated his resignation by getting in a hot-air balloon and heading toward the residence of the chancellor, blowing melodies on a recorder as he wafted over the house.

Next stop was Silicon Valley, where he wrote documentation for the booming personal computer industry. He also did some reporting for a chip-crazy publication called *Doctor Dobbs' Journal;* his bio said, in part, that Raskin "is well known for his heretical belief that people are more important than computers, and that computer systems should be designed to alleviate human frailties, rather than have the human succumb to the needs of the machine." One day in 1976 Raskin found himself in a garage in Los Altos that was the headquarters for a humble new start-up company. It was even a humble garage. There was a workbench along the wall and parts all over the place. Sitting on a stool was a husky bearded guy named Steve Wozniak. And then in walked Steve Jobs.

Raskin, at thirty-three years, was their senior in both age and experience.

"They wanted me to write a manual for the Apple II," Raskin told me. "I was talking fifty dollars a page. They talked fifty dollars for the whole manual." Still, he agreed to the Steves' terms, and they presented him with an Apple II, serial number two. Raskin wrote a literate manual that became a standard for the young industry. He accepted the job of Apple's director of publications—the company's thirty-first employee—officially joining the company, now formally incorporated, in 1978. He insisted on a clause in his contract assuring him that his duties would not interfere with rehearsals for the opera company.

From the first, Raskin believed he knew more than his boss, Steve Jobs. The personal computer industry had its roots in hobbyism—wireheads and amateurs who were fascinated with the workings of machines. Apple was certainly no exception—in fact, as Raskin later told an interviewer, he purposely concealed from his new employers the fact that he held a degree in computer science. "If they had known . . ." he said, "they might not have let me in the company, because there was such an antiacademic bias in the early Apple days."

Wozniak was certainly in the seat-of-the-pants engineering mold. And Steve Jobs? As far as Raskin was concerned, Jobs was no visionary, certainly not a skillful engineer, but a college dropout with an ego problem, a sponge who absorbed the ideas of others. Raskin, on the other hand, had a firm idea of what personal computers could be. They could enable people—*if* the people who designed them made intelligent choices. Before the hobbyists even integrated the word into their lexicon, Raskin was a student of interface. And he was hyperaware that in terms of interacting with human beings, computers were woefully inadequate. He would say it bluntly: "I think personal computers are a pain in the neck."

At first, Macintosh was Jef Raskin's baby. He even named it. Raskin believed that bestowing a woman's name on a computer was a sexist act, a belief he freely shared with others. The Macintosh apple was Raskin's favorite, and thus worthy of being a working code name. Ultimately, after some deft negotiating with the McDonald's fast-food chain on the status of trade names prefixed "Mac," it became the computer's true name.

The Macintosh was Raskin's reaction to another of Apple's mistakes, which, he thought, could have been prevented had the company listened to him: the overblown and overpriced capabilities of Lisa itself. But Raskin had himself to blame for the direction Lisa took.

While a scholar at Stanford's AI lab, Raskin had spent

some time hanging out at PARC, and was impressed with some aspects of the Alto interface. Two things about the Alto struck Raskin as brilliant. One was its bit-mapped display, allowing for the flexibility of type and graphics that one gets from the printed page. The other was its lack of modes. In early 1979, Raskin was working with the Lisa group—he had, in fact, recruited one of his former students, Bill Atkinson, as its star programmer. (At UC San Diego, Atkinson had been a participant in guerrilla theater productions organized by Raskin.) As Raskin tells it, he was the one who first urged that the Lisa engineers go to PARC to see the advanced display. Eventually, the notorious deal was cut and Raskin accompanied the group on its historic visit. Before leaving PARC, he told Larry Tesler, "We don't need this, but I'm glad they saw it." But he had underestimated the effect of Alto and Smalltalk on his colleagues, who were seriously blown away by the digital pyrotechnics.

"As soon as that happened, I was dropped from Lisa," said Raskin of the PARC visit. The next sentence was implicit: *Steve Jobs drove me out.*

As far as Raskin was concerned, it was just as well. As a result of bulking up to run bit-mapped windows and multitasked programs, Lisa was destined, in Raskin's view, to become too large and too expensive. Instead of bringing groundbreaking technology to Apple's core constituency—the people—it was destined to compete with dinosaurs like the word processing systems of Wang or Lanier. "If I wanted to work for a business company, I'd join IBM," Raskin had complained to Apple's president Mike Scott. *A computer can only be successful if it's accessible,* Raskin believed.

Raskin's idea was a computer that was *really* low cost. Something that would not be very powerful, but would leverage its power so that truly useful tasks could be performed elegantly and efficiently. It would be a sharpshooter to the Lisa's bazookateer. Of course, Raskin did not use a munitions metaphor. His model was a modest yet essential tool: the Swiss army knife.

According to Raskin, Steve Jobs hated the idea of a simpler machine. So Raskin went to Mike Markkula, who was then chairman of Apple. Markkula, a former Intel executive whose money helped get Apple off the ground, was interested. He asked Raskin what he could design in the way of a $500 machine.

"I said, 'Nothing,' " recalled Raskin. "But for a thousand I could give you something that could be dynamite."

Raskin began to write in a loose-leaf binder he called "The Book of Macintosh." It included design notes, a business plan, marketing ideas, and a philosophy. "The purpose of this design is to create a low-cost portable computer so useful that its owner misses it when it's not around . . . even if the owner is not a computer freak," wrote Raskin in January 1980. A month later he mused, "The personal computer will come of age when it goes the way of the calculator or the telephone, or probably both . . . it will become a nearly indispensable companion."

The machine would be unlike anything Apple had yet produced. A one-piece tool that fit underneath an airline seat. It combined aspects of the PARC philosophy, including the portability of the Dynabook, with the utilitarianism of the Apple II. (But not a mouse—Raskin hated anything that took one's hands off the keyboard.) You would flick it on and it would be ready to use. The software for the only really important things you used a computer for—writing, calculating, files—would be built in. You wouldn't even have to worry about which program you were using: you would simply work, treating the screen like a piece of paper. It would be as friendly to use as a toaster. Mastering Macintosh would be as simple as dropping a couple of pieces of whole wheat into the Sunbeam.

When Raskin was scrawling his manifesto on the relatively hard-to-handle Apple II, most people likened the task of learning a computer to root canal work: though no one wanted to do it, one day they might have to do it. If that unbidden calamity befell them, they would glumly accept what looked to be a hellish process. So what Raskin proposed actually seemed preposterous. The claim that he could deliver this unheard-of conviviality for under $1,000 was even more outrageous. Yet in his exacting, almost pedantic manner, he was convinced that it would happen. And though he was ultimately destined to be a loser in the corporate politics that seethed underneath the surface of Apple Computer's dippy new-age culture, he was canny enough to get his pet project going.

Operating with a degree of stealth so that the vulnerable young project would not draw the dangerous attention of Steve Jobs, Raskin gathered a small team to begin implementing Macintosh. The key hire was Burrell C. Smith, a diminutive blond twenty-two-year-old working in Apple's repair department. Smith had little formal training—"a total of ten courses at Foothill College," as he once described his education in electronics. But when it came to computer hardware Smith was The Natural.

His idol was Steve Wozniak. Wozniak's creations in silicon were legendary; in particular his work on the Apple II floppy disk drive was acclaimed as a masterpiece of technology. When more pedestrian engineers looked at the Woz disk controller, they saw confusing, almost counterintuitive leaps that somehow, almost miraculously, led to high performance. Smith saw a haunting, almost divine logic. Every connection had its reason. It was a world where all was right. Smith felt comfortable in these worlds where logic ruled, and he loved to isolate himself from the messy loose ends of reality and devote all his creative energy to working in that logic-grounded realm. He felt like himself there.

In a way, Smith's stint in Apple's repair department was like Lana Turner's possibly apocryphal presence at Schwab's Drugstore. Both were waiting for stardom to tap them on the shoulder. Both got their wishes.

Bill Atkinson was one who put Smith and Raskin together. He brought the young engineer to Raskin's house one night, and said, "Here's the guy who's going to design your machine for you."

"We'll see," said Raskin. And after taking the measure

of Smith, Raskin agreed. But it was the utterly Macintosh thing to do. At Xerox PARC, and even at the Lisa division at Apple, no one with such a scanty résumé would have been asked to come up with the main logic design of the computer. Smith was elated. "It was the one chance of a lifetime to go through the cracks of the corporate culture," he once told me. "We raced through the elevator door just before it shut."

This was the Macintosh team back then: a scrappy assortment of outcasts and mutts. It was an interesting pedigree. Instead of being handed down to the people from the gods of computer science, Mac was a bottom-up project, generated by the actual hoi polloi. "We design products that we ourselves want to own," wrote Raskin. His challenge was to maintain his vision during the long struggle of product development. But first he needed a prototype.

"It was a statement of simplicity," said Smith of Raskin's vision. "Of absolutely delivering the dream of a machine that you could just turn on—you don't have to have a machine that does everything in the world." It was to be a bit-mapped machine with a built-in display, built-in keyboard, and built-in file storage (in a move of ultra cost consciousness Raskin wanted the already-obsolete cassette-tape storage instead of a floppy disk drive).

Over the 1979 Christmas holiday, Smith virtually lived inside the deserted Apple building, attempting to crank out a system that would fulfill Raskin's wish list. For parts, he scrounged around various offices and supply rooms, scavenging what seemed useful. For the central processing unit (CPU)—the chip that was the center of all computer commands—he used Motorola's 6809, which was no more powerful than the CPU in the Apple II. After a week of work, he managed to build the main logic board. Early in the new year he grabbed a soldering iron and inserted the board in an Apple II box. He finished the job late one night and went home, but before he did, he notified a friend of his in the Apple II division of the project's status. He added an informal challenge—could the programmer get something to display on the screen? Then he would know his prototype worked.

The programmer was Andy Hertzfeld. After the building was abandoned by day-jobbers, Hertzfeld found his way to Burrell's desk—and the prototype. He sat down at it, deciphered Smith's sketchy instructions on how it worked, and spent the entire night trying to conjure a picture on the screen. The next morning, Burrell Smith came in to see that the Macintosh was alive. The image that Hertzfeld chose to display was indicative of the frivolity that would eventually be part of the machine's personality: it was a picture of Scrooge McDuck, underlined with the handwritten greeting, "Hi Burrell!"

Despite the quick beginning, the effort moved slowly for much of 1980. Raskin was proceeding gingerly, aware that Macintosh was a sitting duck for cancellation. At a couple of points that year, the project actually got axed, and Raskin had to beg more indulgence. In September, a new software architect joined the small group, Guy "Bud" Tribble, a lean, tousle-haired programmer who had been teetering between a career in technology or medicine. Tribble had known both Raskin and Atkinson in San Diego, and was well into a program in neurophysiology at the University of Washington that would lead to a Ph.D. and an M.D. But the prospect of denting the universe drew him to Macintosh, and he took a leave from Washington to work at Apple. His dissertation on feline neural disorders would wait.

That fall, the group moved out of the main Apple building into an office on Stevens Creek Boulevard, referred to by the Mac people as "The Good Earth" after the faux health food restaurant that stood in front. The five-person group shared a single room, with a small anteroom to the side. In the center of the room sat a stack of cardboard boxes that had once held equipment. This was used as sort of a playroom. At the drop of a hat, everyone would stop work and throw Nerf darts at the boxes, or engage in Frisbee matches. In the name of encouraging creativity, Raskin blessed these outbursts.

The project was canceled, again, in October. Raskin begged Apple president Mike Scott for a three-month reprieve. The team was still hoping to come up with something impressive enough to extend the project. Complicating matters was the progress of the Lisa. Apple was pouring millions of dollars into its design; in the wake of the PARC visit late the previous year, Lisa was firmly on course to offer the first graphical user interface for a personal computer. Raskin, of course, would not consider something similar for his ultrasimple, ultra-cheap Macintosh. But others in the company were so enamored of the idea that they urged the Macintosh team to move in that direction. The first step would be the adoption of the same CPU chip that the Lisa used, the considerably more expensive Motorola 68000 processor. Bud Tribble, who was aware of the fantastic things his friend Bill Atkinson was writing for the Lisa team, urged that the Macintosh team follow suit. Tribble knew that, to some degree, he was betraying Raskin with this position, but he really believed it was best for the project. Ultimately, adopting Lisa technology would make Macintosh more the computer that *he* would like to own.

But the chief proponent of this shift was Steve Jobs, whose disruptiveness had emboldened the engineers and executives on the Lisa project to bounce him off, with the blessings of Apple's chairman and president. Despite Raskin's efforts, Jobs came across the Book of Macintosh, and was so impressed with Raskin's vision about a computer being as easy to use as a home appliance that it became part of Jobs's standard spiel for years thereafter. Jobs began to insinuate himself into the skunkworks project behind the Good Earth, and Raskin's pure vision was as good as gone.

"It was clear that Macintosh was the most interesting thing at Apple—and Steve Jobs took it over," said Raskin.

The takeover proceeded by increments. Steve Jobs was not a technical wizard, but he thoroughly understood the mindset of the people who were. So when he tossed a challenge to Burrell Smith, daring the young engineer to design a prototype Mac for the 68000 chip, the job was a cinch—despite knowing Raskin's mania for a low-price machine, Smith rose to the bait, and spent all of December 1980 to get it working. Raskin was not happy about the development, but Jobs outranked him. All Raskin could do was quickly recalculate the rock-bottom cost of manufacturing this new version of Macintosh.

Smith got the prototype working by Christmas—almost a year to the day after his first effort. This was a remarkable enough achievement but even more impressive were some of the tricks he used in the process. Back at Apple's main buildings no less than twenty-four high-priced engineers building the Lisa had taken two years to construct a carefully planned architecture around the 68000 chip—everything carefully charted and approved by committee. Smith, a true hacker and packrat, spun his own version out with off-the-shelf parts and magic. It was not easy—fitting the mighty Motorola 68000 to the rather modest framework of the Macintosh electronics was like strapping a jet engine to a Honda. Yet Burrell, using a technique called "bus multiplexing" found a way to pull off the trick elegantly—he was able to draw the benefits of the microprocessor's power without overwhelming the Macintosh hardware. The result was that Smith's computer was more eccentric, but ran twice as fast as the Lisa.

This was all Steve Jobs needed to hear. The Macintosh, a computer designed to be as easy to use as an appliance, a computer that would cost thousands of dollars less than the Lisa (and even cost less than the comparatively dinosauric Apple II), could outrun the Lisa itself! Certainly a measure of revenge would belong to Jobs if he headed a design team that trumped the project from which he had been humiliatingly bounced. But there was something else at work. Steve Jobs was increasingly known as a national figure, the exemplar of the new age of entrepreneurism. Yet for years, he hated the fact that everyone in Silicon Valley considered the Apple II, the company's only truly successful product, to be solely the creation of his partner Steve Wozniak. (This despite Jobs's brilliant idea that the computer should be packaged not like a wirehead hobby project, but a consumer product.) Jobs instead was viewed as sort of a hustler, a slick marketer. This was a continual insult to Jobs, who yearned for respect in every way. Now he had the means to earn it: he would be the official leader of the Macintosh project.

"When Steve started coming over, Jef's dream was shattered on the spot," Joanna Hoffman, the lapsed archaeologist whom Raskin had hired in November as the first marketing person on the team, later told me. "It was difficult on everybody and there was an allegiance to Jef, but Steve had his own compelling aura. He immediately started talking about what it would look like, feel like, how we would sell it . . ."

For the first couple of months in 1981, Raskin and Jobs tried to work together. But already, two visions were clashing. Steve Jobs kept pushing for the Macintosh to be a smaller, affordable, sleeker version of Lisa—

something so "insanely great" that it would actually bury the corporate computer that his company was spending $50 million to develop. Raskin kept fighting to keep costs lower, to keep focus tighter, to keep the goals more modest—basically to stave off the creeping Lisa-ism.

Jobs immediately moved to put his mark on the Macintosh team itself. His new recruits were the people who had been the creative forces behind the Apple II: Rod Holt, the analog engineer; Jerry Manock, who had designed the now-familiar plastic casing; Randy Wigginton, who worked on the software; and even Steve Wozniak himself. The Macintosh project moved from the Good Earth to another location a few blocks from headquarters, nicknamed Texaco Towers after a nearby gas station.

Two leaders was one too many.

Jobs claimed that Raskin's pace was too leisurely—at the rate he was going there would never be a Macintosh. Jobs claimed that he could motivate the team to complete the project in little more than a year.

Raskin, meanwhile, insisted that Jobs was incapable of running a successful project. According to Raskin, Jobs simply would not listen to people, would not consider alternatives, could not make rational judgments. His style was imperious and overbearing. What had ruined Steve Jobs was his success—before even turning thirty he had achieved fame and riches exceeding that of some truly great industrialists. "How can you believe any criticism when everything you do turns to gold?" asked Raskin. If Jobs were truly a genius or visionary, this wouldn't have been so bad, but Raskin thought Jobs really didn't know very much. How could a person be a leader if he constantly bludgeoned people on the team without even holding a solid idea of what should be done? An Apple vice president had once expressed a sentiment that Raskin considered accurate: "You can't use the words Steve Jobs and manager in the same sentence."

But it was Raskin himself who exercised that linguistic oxymoron, in a memo to Apple president Mike Scott on February 19, 1981. "While Mr. Jobs's stated positions on management techniques are all quite noble and worthy, in practice he is a dreadful manager," wrote Raskin. By then the game was just about over—Jobs had informed Raskin that his duties thereafter on the project he initiated should be limited to those of publications director. Raskin's response was to enumerate all of Jobs's failings as a project leader: "He was late for appointments, he attacked people's work without understanding it, he sowed divisiveness and discontent, he played favorites, he had no idea of realistic scheduling . . ."

The memo was Raskin's burned bridge. He later claimed that it represented the sense of the entire team. "We were going to do it as a group, but others chickened out," he said. "And I was the only person who stood up to him." Thereafter, Raskin was off the project and Steve Jobs was solely in charge. Officially Apple's top officers had granted Raskin some time off from the project, but as it turned out, that was the end of his tenure at Apple. He quit within months. Some cynics thought that Markkula and Scott allowed Jobs to have his way only because the Macintosh was a backwater, isolated from the company not only by geography, but in direction as well—a product deemed unlikely to succeed.

In retrospect the infighting between Raskin and Jobs was much more than a fleeting clash in the business world. The pushing and pulling had its effect on the finished product—which in turn affected the lives of literally millions of people.

It was Raskin who provided the powerful vision of a computer whose legacy would be low cost, high utility, and a groundbreaking friendliness. Raskin was so protective of these qualities that he fought even the most obvious concessions to granting the machine more power. The processor he chose as the brain of his machine was clearly too feeble. The amount of digital memory he wanted was absurdly inadequate. (Even though Jobs's version of the computer held twice as much memory, it was still woefully sparse.) And his resistance to some of the PARC innovations embodied in Lisa were as much based on pedantry as on a sure evaluation of what would make a computer easy to use.

Jobs entered from the other extreme. As soon as he chose to be part of the Macintosh project, he demanded that the Mac take on a mouse, just like Lisa. Raskin was aghast—in his mind the mouse was nothing more than a nuisance. Raskin believed that Jobs really didn't know what he was talking about—"His conception of the Macintosh was fluid, it kept changing," he complained to me soon before the Mac launch. "The last person who gets Steve Jobs's ear is the one he believes." Yet in the case of the mouse it was Jobs who was correct. Not only is a Macintosh without a mouse now unthinkable, but most people agree that any personal computer is incomplete without a mouse, or at least a pointing device to perform the same job.

A few years after leaving Apple, Raskin finally got the opportunity to put his own mouseless vision in play, starting a company called, significantly, Information Appliance. It made an extremely easy to use computer eventually marketed as the Canon Cat. It made little impact on the computing world, and no dents whatsoever in the universe.

The upheaval at Macintosh was only one of several fateful events occurring at Apple in February 1981. Early in the month, Steve Wozniak crashed his small plane on a runway in Berkeley. For a time, he had amnesia. Eventually he would recover, but he never returned to the Macintosh project. Apple itself would suffer dramatically from his loss. Woz never was a force in management but his mere presence had been a palpable asset: he was, after all, the embodiment of the original garage engineering feat. As Jobs could bitterly testify, Woz *was* the Apple II, and the Apple II was still by far the bulwark of Apple's revenues.

Apple's recent sequel to that computer, the Apple III, was a flop. It was not a *bad* computer, and indeed, it was a somewhat better machine than its main competition, Big Blue's IBM PC. But it was boring. Randy Wigginton, who at age sixteen was one of Apple's founding employees, once told me, "The Apple III was kind of like a baby conceived during a group orgy, and [later] everybody had this bad headache and there's this bastard child, and everyone says, 'It's not mine.' " People hoped for more than incremental improvements from Apple; at the very least they expected something that would make them abandon their Apple II's and hunger for the successor. Worse, Apple shipped the first fourteen thousand or so III's with a horrid hardware flaw that led to an abundance of dead computers and enraged customers. The embarrassed company recalled all the early units, and fixed the problem, but word spread that the machine was unreliable. No matter what Apple did, no matter how it improved the Apple III from then on, the computer had the smell of death to it. Apple supported it for a couple of years, and then quietly pulled the plug. Its main legacy was to dampen the sense of invincibility that had arisen at Apple.

Meanwhile, IBM's PC had finally provided an excuse for corporate management information services executives—most of whom had regarded the personal computer movement with a measure of anxiety and reserve—to begin purchasing the things. The PC was a conservative computer, crafted with a degree of openness hitherto unexercised by the behemoth known as Big Blue. Yet its interface, if you could call it that, was squarely of the old school—a command-line operating system that failed to make computers one iota friendlier than before, Charlie Chaplin commercials notwithstanding. Apple's response at first was a sigh of relief—rumors had abounded that IBM would make a technical leap that would signify its PC's were a clear step beyond Apple's. So confident was Apple that it took space in the *Wall Street Journal* to greet its competitor. "Welcome IBM," read the ad. "Seriously." As the IBM PC came to dominate the industry, these words would haunt Apple.

Bill Gates once told me that he was visiting Apple on the day that IBM introduced the PC. "They didn't seem to care!" he said. "It took them a year to realize what happened."

"When we started the Macintosh project IBM didn't have a machine," said Chris Espinosa, a veteran of the Apple II garage who assumed the Macintosh publications job after Jef Raskin left. "We looked very carefully [when the PC came out], and at first it was embarrassing how bad their machine was—a half-assed, hackneyed attempt at the old technology. Then we were horrified [at its success]."

But for the first time in its brief, shining history, Apple had reason to worry. Not only was its future in peril, but current revenues could not support the growth it had undertaken. In late February of 1981, a date thereafter known in Cupertino as "Black Wednesday," Apple laid off forty-one workers. In an era where corporations vied for meanness and leanness the cutback was by no means exceptional. Years later people would point to splashier, more public events to mark the loss of Apple's virginity. But those at Apple during Black Wednesday still will contend, with quavering voices, that it truly marked the end of innocence.

There was only one outpost in the culture where the Spirit of the Garage lived, and that was at Texaco Towers, where people were still reeling at the loss of their project's founder: the Macintosh project. If it wasn't clear then, it soon would be—Apple's future depended upon it.

6

The man who programmed the very heart of the Macintosh, the code residing in a computer chip (called ROM, an acronym for Read Only Memory), believed that his calling might well have been that of a novelist. A visitor to the cluttered Palo Alto bungalow he lived in during the 1980s was just as likely to be engaged in impassioned conversation about James Joyce as he was to a discussion about new system software. Andy Hertzfeld was walking disproof of the stereotype that computer geniuses were narrowly focused and nerdy: besides literature, he followed pop music closely and religiously attended 49ers games. In certain other ways, though, he was a classic computer hacker—he loved to amaze people with startling stunts requiring technical virtuosity, joyously treading on the foul-lines of possibility. And he despised authority.

Largely through Andy Hertzfeld, both of those traits would become part of the personality of the Macintosh.

A native of the Philadelphia suburbs, Andy became interested in computers during high school, and quickly discovered that he had an exceptional talent for working with them. This did not preclude his ambition from occasionally surpassing his reach. One of his first programs was designed to match partners for a school dance; unfortunately, when Andy's creation was put to use one girl became unexpectedly popular, finding herself matched to every boy in the room.

After studying math, physics, and computer science at Brown, Andy entered UC Berkeley as a graduate student in the latter field. The coursework was rigorous but boring. He began to fear that the career he had chosen would be equally dry. In his nightmares, he churned out workmanlike code for creepy bosses in suits. Then he discovered the Apple II. "It changed my life," Andy told me on that first day we met. "The more I learned about it, the more I was impressed with its brilliance." He dropped out of graduate school and began writing Apple programs. One of his hacks filled a gap in the Apple II that Jef Raskin had first identified: it displayed only uppercase letters. His first impulse was to give the program away—in Andy Hertzfeld's mind, anything that helps people use a computer more efficiently is a good in and of itself. But a friend convinced him to sell it, and Hertzfeld made $40,000 in a few months.

Andy went to work for Apple in 1979. In some ways it was a dream; he had access to the secrets of the Apple II, and even began a friendship with his hero, Steve Wozniak. On the other hand, the company was just beginning its accommodation with hypergrowth, with some disturbing side effects. A year after Hertzfeld arrived, Apple went public, and for months thereafter many of the instant millionaires at Apple were obsessed with the daily stock price.

This unseemly behavior dismayed Andy, but he was downright depressed at the continuing influx of "bozos" from established companies like H-P, or even IBM. Their values clashed with the hobbyists and hackers who formerly ruled at Apple. The newcomers seemed more concerned about achieving competence than (as Alan Kay might say) "dealing lightning." The personal computer world was Andy Hertzfeld's analog to never-never land, an alternate universe where one could make a living without ever growing up; in fact, retaining one's childlike wonder was the key to the sort of engineering that Andy specialized in. Good software not only did a job, it made your jaw drop—*how did he do that?*—or triggered a delighted grin—*that is unspeakably* neat. Hertzfeld was driven to write programs that evoked the same exhilarating vertigo as a Disney "E" Ride. Watching a Hertzfeld demo was sort of a litmus test for bozohood: losers and suits wouldn't get it.

The final straw almost came for Andy on Black Wednesday, when Apple's president Mike Scott laid off the forty-one workers. (A few months later Scott himself received a pink slip.) Hertzfeld was in tears; finally his illusion that Apple was anything but another heartless corporation was shattered. The only way he would stay at Apple, he decided, was if he could work on the one project where the original spirit of Apple still lived: Macintosh. He had lusted for this since the Scrooge McDuck demo, but Jef Raskin considered Andy too unprofessional, a seat-of-the-pants hacker. Hertzfeld explained his preference to Scott and before the day was out, Steve Jobs appeared in his cubicle. "Andy," he said, "You're working at Macintosh now." Jobs then scooped up Hertzfeld's equipment and led his new acolyte to the car. At Texaco Towers, Hertzfeld saw that the desk to which he was assigned still held the paraphernalia of its previous owner—Jef Raskin.

Andy first performed the technically challenging task of getting the Apple II disk drive to work on the Macintosh. But as he became more integrated into the Macintosh team—at a crucial time, because Bud Tribble was out sick for over a month with meningitis—the task of writing the Macintosh's built-in software fell to him. In a sense these ones and zeros—permanently embodied in a ROM chip that would sit on the logic board—were the Macintosh's DNA. Future versions would incorporate intentional mutations—improvements—of this blueprint, but essentially the personality of the Macintosh was embodied in the ROM.

So compact and ingenious was Hertzfeld's programming that industrial pirates attempting to copy the ROM by "reverse engineering"—getting a printout of all the bits and trying to decipher it—would indeed be faced with a task similar to that of molecular biologists attempting to crack the genetic code.

What was in the ROM? The Macintosh toolbox. This began with QuickDraw and extended to things with names like the Window Manager, the File System, the Resource Manager, the Font Manager and TextEdit—in short, it was an engine that, when properly connected to an accommodating software application, churned out what would become the unmistakable Macintosh interface.

The toolbox turned out to be one of the most important reasons for Macintosh's success. By modifying the PARC interface paradigm, Apple was venturing into unknown territory: if all went well, all previous personal computer software would soon be obsolete. Applications software that did exploit the virtues of the new interface would be a clear leap forward—but in 1982 this software did not exist. Definitely a problem.

To solve it, Apple could have adopted one of two approaches: create the software itself, or recruit other companies to do it. The Lisa team embraced the former solution. Lisa's engineers wanted the advantage of having a complete ensemble of software available for the machine, right from the start. From inside Apple came LisaWrite, LisaDraw, LisaConnect, and LisaProject. The very existence of these programs ensured that independent software developers would have little incentive to produce Lisa software: buyers assumed that the proper tools for Lisa would be those created by the company that made the computer itself. These authorized products would be sold at the same places the computer would be sold. Who could compete against that?

Macintosh took the other path. Apple intended to inspire outsiders to devote their resources—intellectual and commercial—to creating a software library for Macintosh. These other companies would be referred to as third parties, welcome interlopers in the relationship between Apple and its customers. Their role would be similar to, and as crucial as, the symbiotic third party of bees, whose business of pollination is essential in the relationship between the parents of prospective plants.

Apple had seen how a thriving third-party market could bolster a computer's fortunes—this had occurred with the Apple II, which boasted a software library of thousands of programs, very few of which were developed by Apple. Independent developers had written dozens of word processors, and even programs for obscure functions like cattle management and tarot card reading. Thousands of business people bought Apple II's solely because it was the only computer that ran VisiCalc, published by a third-party developer. The proper lesson from all this was that personal computer companies are just as well off letting others produce great software.

On the other hand, the aggregation of software available for the Apple II was just that—a motley. Though each program had to labor under the burden of the loathsome command-line interface, some applications were easier to use than others. But the features from the best programs were seldom if ever present in others—each program tackled interface problems in pretty much its own way. A keyboard combination that saved a file in one program might delete a word in another. You couldn't even quit a program in a standard way. Every time a user purchased a new program, he or she had to scale a fresh learning curve.

Macintosh would change that. The Lisa interface would be adapted and taken a step further. In the same way that Bill Atkinson became the key interface person on the Lisa—he wrote the software that controlled the screen—Andy Hertzfeld took on this role for Macintosh. Since Macintosh was a smaller project than Lisa, with almost no bureaucracy, Hertzfeld had license to be looser. "My job was to decide what of Lisa to keep," he later explained. "I did it my way, following my intuition. The idea was to make the interface reach a lot more people. We had a drive to be simpler—we had only a fraction of the Lisa's hardware resources." And since Andy admired Atkinson's work, and disdained most of the other Lisa engineers', "Anything Bill Atkinson did, I took, and nothing else."

Hertzfeld had another guiding principle for his decisions—the spirit of the Apple II. In his view, Macintosh was a reincarnation of the Apple II. He wanted a computer that could generate the same excitement, the same irreverence, the same overabundance of creativity as that machine had, but for a far wider audience, the masses of people who were not normally drawn to computers.

In contrast to the Lisa approach, the Macintosh team implemented its interface issues on an ad hoc basis. When disagreements came, the criteria was not What would an average Joe like? but What do *we* think is the right thing? Disputes would often be settled by a vote. The bias was almost always toward the option that would be more amusing to the user, or simply seemed to have the flavor of what was becoming Macintosh. There were differences in the scroll bars on the side of the windows, on the title bars on top of the windows, even the addition of a three-dimensional shading effect on the pull-down menus. There was the more significant variation of allowing folders to reside within folders, ad infinitum. There was the innovation of the "Apple menu," a command on the left of the menu bar that would pull down a set of tools called "desk accessories" that would always be available—things like a calculator, a clock, and even a little puzzle. These were all innovations that pleased the Mac team, features they wanted on their own dream computers. Though this design approach was less scientific than that used at PARC or Lisa, it was arguably superior. Cumulatively, it resulted in an interface that not only presented a coherent physiognomy to a user, but a rather fascinating one.

Not until January 1982 did the Macintosh team finally sit down to formalize things. After a week of discussion Joanna Hoffman was put in charge of creating a document called "Human Interface Guidelines" that codified what came to be referred to loosely as "Macintosh religion." Some people would later use the term derisively. But the religion assured that the look of the screen would be the same no matter what application program was running at a given time. The parameters for operations such as dragging text, selecting objects, and quitting programs would be consistent and predictable. All the dialog boxes—those little windows that come on the screen at decision points and ask you to signal your choice by moving the cursor over a button—would look alike. Every program would have similarities on its menu bars, and similar commands. When one used a Macintosh application, he or she could know with virtual certainty that saving the document could be accomplished by invoking the drop-down menu under

FILE and moving down to the fourth item, SAVE. (Or using the keyboard combination COMMAND-S.) Copying text was the fourth item down under the EDIT menu, or COMMAND-C. The undo function—a valuable innovation in and of itself—was the first item in that menu, or COMMAND-Z.

For software developers, these guidelines would be ignored at one's own risk. Once users became accustomed to the standard conventions of Macintosh computing, they would reject applications that flouted those standards. It was as if users had developed an immune system that resisted outsiders not tagged with the Macintosh imprimatur. At least that was the hope. Meanwhile, the Macintosh toolbox acted as sort of a built-in ease-of-use czar. Sitting mainly in Andy Hertzfeld's ROM chip, this toolbox performed like a telephone switchboard, accepting input from the application and placing a "call" to a specific aspect of Macintosh. One call might evoke a window, with the standard scroll bars and title bar. Another might control a menu. Another might trigger a dialog box, offering the user one of several command options.

As a result, the entire software base of Macintosh became a coherently created world in itself, one with an immediate familiarity to anyone who had mastered the elemental skills of using the machine. (And these skills, being visually clear and intuitively logical, would be a cinch to learn in the first place.) You could launch a strange application, and accomplish something instantly, without even touching the manual. After some painless exploration, and perhaps a glance or two at the documentation, you could probably get serious work done.

It was an ambitious plan, and, amazingly, it worked. It also set an implicit example for others, in computer design and elsewhere. Some years later, after praising the effect of the toolbox, design critic Donald Norman wrote, "Now if we can enlarge a similar spirit of standardization to the machines of all manufacturers, all over the world, we would have a major breakthrough in usability."

Soon after Andy Hertzfeld joined the Mac team, Steve Jobs literally changed the shape of the machine. Steve Jobs considered the look of his products to be paramount, and his intentions with the Macintosh design were particularly ambitious. In addition to fulfilling all the functional requirements—portability, ease in setup, and ergonomic thoughtfulness—Macintosh had to satisfy two additional criteria. First, it had to be a physical statement that this computer, unlike any that came before it, was easy enough for anyone to use—fun, even. And second, perhaps even more dear to Jobs's heart, the Macintosh had to be a gorgeous object in and of itself. Jerry Manock, the industrial designer who crafted the casing for the Apple II, was given the similar assignment for Mac. "If it would get into the Museum of Modern Art design collection, Steve would be very happy," he told me.

Elegance was a mania for Jobs. "It goes back to the

first brochure we ever did at Apple," he said. "It was white, with a picture of an apple. Fruit, an apple . . . that simplicity is the ultimate sophistication. What we meant by that was when you start looking at a problem and it seems really simple, with simple solutions, you don't really understand the complexity of the problem. Your solutions are way over-simplified. Then you get into the problem, and you see that it's really complicated, and you come up with all these convoluted solutions. That's sort of the middle, and that's where most people stop, and the solutions tend to work for a while. But the really great person will keep on going and find the key, the underlying principle of the problem. And come up with an elegant, really beautiful solution that works. That's what we wanted to do with Mac."

The Macintosh had to be beautiful in *every* way—even the parts that no one but service technicians would ever lay eyes on. When Burrell Smith delivered the prototype for the first Macintosh circuit board, Jobs rejected it on aesthetic grounds. He once explained this to an interviewer: "When you're a carpenter making a beautiful chest of drawers, you're not going to use a piece of plywood on the back, even though it faces the wall and nobody will ever see it. You'll know it's there, so you're going to use a beautiful piece of wood on the back. For you to sleep well at night, the aesthetic, the quality, has to be carried all the way through."

This is not to say that Jobs was consistent in his demands. A case in point was the so-called Cuisinart Mac, when Jobs became temporarily obsessed with the boxy-looking machine that made cole slaw, juilienned carrots, and kneaded bread dough. "It was a two-week exercise," explained Jerry Manock. "Steve would go to Macy's for four hours, looking at food processors."

When Manock had his finished version, however, it was worth all the trouble. The Macintosh casing was so distinctive that its visual presence would become as recognizable as a Volkswagen bug. Unlike virtually every previous computer, whose orientation was horizontal—a flat, typewriter style keyboard built into typewriter-style box on which a monitor sat—the Macintosh was vertically oriented. A main body with a small monitor sitting above the main workings of the machine, all encased in plastic. A detachable keyboard completed the ensemble. The entire structure was to be extremely compact; the "footprint" of a Macintosh on one's desk would be comparable to a flat piece of paper. It stood only fourteen inches high.

There was something lovable about that shape. It was . . . the Macintosh.

Working for Steve Jobs was a decidedly mixed blessing. On one hand, he was the most passionate leader one could hope for, a motivating force without parallel. Tom Sawyer could have picked up tricks from Steve Jobs. Time after time, he insisted that the Macintosh was going to shock the world, be not merely great but insanely great, and it was clear that he believed it. Sometimes the wizards of Mac would roll their eyes at his rants—remember, they referred to Steve as a walking Reality Distortion Field—but they were flattered, too, and determined to transform the hyperbole into truth. One day, for instance, they were trying to "bring up" the main logic board—soldering the chips to the custom-made circuitry—and Jobs challenged them: If you finish it by midnight, we'll all go out for pineapple pizza at Frankie, Johnnie, and Luigi's! (Burrell's penchant for pineapple pizza had infected the whole crew.) And they did.

But every subordinate of Steve Jobs also saw his down side. He wore his demons on his sleeve, and was sloppy in dealing with them. Jobs's criticism took the form of acid humiliation, even on occasions when Jobs himself was unqualified to judge the quality of the work in question. You might work on something for a week, when Jobs, exercising what Jef Raskin once called MBWA (Management By Walking Around), would chance upon your cubicle, give a quick glance, and declare, "This sucks!" One new Macintosh worker received a visit from Jobs on his first day: "I want you to know you've really blown it," Jobs began the conversation, complaining about a preexisting problem that the employee hadn't even addressed yet. Defending yourself was out of the question. You certainly didn't want to cross him—once someone fell out of Jobs's favor, it was as if he or she no longer existed in his mind.

Why did they put up with it? Why did they work hundred-hour weeks (many for salaries of no more than $30,000) producing first-class work and receiving the worst sort of abuse? Apparently even the darkest side of Steve Jobs, a view apparent all too often, was not so imposing that it eclipsed his charisma. His charm was powerful mainly because it was reflective: working for Steve Jobs was not so much being in his service as it was sharing a special dream, a dream he managed to evoke in breathtaking hues. (It did not hurt that he had already realized more dreams in his brief existence than almost anyone had in longer lifetimes.) His exacting standards, even when misguided, reminded the Mac people that they were not salary workers but revolutionaries on a mission. When Apple's new president, John Sculley, came across the Macintosh people, he realized that he was witnessing a phenomenon unknown in the behaviors of what he had known as a highly motivated soft-drink marketing team. Those people were driven, mostly by personal ambition, to beat Coca-Cola. But the Mac team acted as if on a mission from God. Sculley later tried to describe it in his memoirs: "It was almost as if there were magnetic fields, some spiritual force, mesmerizing people. Their eyes were just dazed. Excitement showed on everyone's face. It was nearly a cult environment."

Sculley was not the only one who considered the Macintosh team cultish and fanatic. Steve Jobs had taken pains to encourage the group to think of itself as a class apart from the rest of the company, an exalted duchy within the Cupertino campus. "It felt like very early Apple—it felt like the garage again," Chris Espinosa, a veteran of the Apple II days, told me. "The same juxtaposition of cheap surroundings and expensive equipment." And the same feeling that they were in on the ground floor of history.

In a sense, Bandley 3 housed a miniature Manhattan

Project, a secret initiative devoted to producing a devastating technology. Though they did not maintain secrecy with the same mania as the Los Alamos scientists—at one point Steve Jobs showed a prototype Macintosh to Joan Baez, whom he'd been dating—the Mac designers did manage to keep quiet with uncharacteristic resolve.

One of Jobs's slogans proclaimed, "IT'S BETTER TO BE A PIRATE THAN JOIN THE NAVY." Forget that they were employees of a billion-dollar corporation—the Mac team was a raucous band of buccaneers, answering to no one but their Captain! This conceit reached its apogee when the Mac designers actually flew a skull-and-crossbones above Bandley 3. (The eye of the skull was replaced with the Apple logo.)

Calling the Mac team a cult, however, unfairly characterizes their labors. Implied by that appellation is that somehow they had imbibed some sort of groupthink Kool-Aid. That was not the case at all. It was not blind faith that drove them to change the world, but a belief very well grounded in reality. Day by day the evidence accumulated that they had it within their power to create something a quantum leap better than anything the industry, indeed the world, had ever witnessed. They understood that if they surpassed their best—gave not merely an honest effort, but one tinged with sweat and sacrifice—their artistry indeed could make a dent in the universe.

"A lot of times people don't do great things because great things really aren't expected of them and because nobody really demands they try and nobody says, 'Hey, that's the culture here, to do great things,' " Jobs explained to me. "The environment we set up at Mac assumes that this special, hand-picked team is the best in the world at what they do—there is none better. And being a pirate means really going beyond what anyone thought possible—a small band of people doing some great work, really great work that will go down in history. Rather than joining an organization, where there's a lot more process, many more layers, and more of a guarantee you'll make something good, but almost a guarantee that it won't be great. It means you can fail, but because you're really great you're willing to take on that risk."

Bill Atkinson understood Jobs's methodology as a Darwinian principle that led to insane greatness. "Either people grow into [the pressure] and become great, or they go down," he told me. "You only get one chance to change the world. Nothing else matters as much—you'll have another chance to have vacations, have kids."

Jobs literally made them feel that the quality of Macintosh was a life-and-death issue. At one point in the software development process he was worried about the machine's distressingly lengthy start-up time—almost thirty seconds. He zeroed in on system software programmer Larry Kenyon, who believed he had already squeezed as much speed as possible from the machinery. "Even if it took you three days to make it a *single second* faster, it would be worth it," Jobs hounded him. "If ten million people use the computer, in one year alone, that's about 360 million turn-ons. How many lifetimes does 360 million seconds equal? Fifty? *Would you take three days to save fifty people's lives?*" Kenyon wound up shaving not one but three seconds off the start-up time, sparing a hundred extra souls from the Reaper.

Given the fact that Jobs was not an accomplished engineer, not an ergonomicist, not a trained visual designer, he was nonetheless correct on a startling number of issues. Atkinson once recalled to me a time when he was debating which "primitives" to include among the standard shapes that QuickDraw could generate. He knew he would include an ellipse, a rectangle, and a circle, but decided against including a "round-rect," a shape similar to a rectangle but with rounded corners instead of squared ones. Jobs vociferously disagreed with him about the omission. To convince Atkinson, he dragged him out of the building and walked him around the parking lot, identifying a surprising number of rounded-corner rectangles—the curbs, the NO PARKING signs. Eventually Atkinson came to understand that the shape was vastly underrated. He made the round-rect a QuickDraw primitive, and never regretted it. You can see this round-rect now on every dialog box and button on the Mac.

That was a case when Jobs's vision was on the money. But his obstinence occasionally led him to demand the wrong thing. Sometimes he was so obviously mistaken that his subordinates conducted silent mutinies. One example concerned the Macintosh's memory. Memory size helps dictate a computer's capabilities: the larger the memory, the more complicated tasks it can handle. But though prices were always falling, memory was expensive. Too much memory, and the computer will outprice itself out of existence. Memory is measured in kilobytes, bytes being the equivalent of words in computerspeak. The Apple II had 48K bytes of memory. A standard 1982 IBM PC shipped with 64K—the same memory size that Jef Raskin had originally envisioned for Macintosh.

But once Jobs decided that the Macintosh would have a graphic interface that limitation became impossible. He dictated a memory of twice that size, 128K. His engineers soon realized that this was almost as ludicrous. As the people at PARC well realized, a graphical interface requires much more memory than the previous standard. The Lisa designers also learned this the hard way, and their computer shipped with 1024K, or a megabyte of memory. The Mac team, with typical immodesty, figured they could present a flashier system than Lisa with half the memory—but not a tenth! They argued for more, but Jobs was insistent. So they proceeded to compress a thousand clowns into a Random Access Memory version of a Volkswagen. But not without secretly implementing a scheme whereby the computer could also work with a bigger load—512K. When the price of memory came down—the soothsayers of Santa Clara County were saying this was only a matter of months—users could upgrade the computers to bigger memory chips, and new Macintoshes would have more formidable memories.

Another invisible rebellion was launched on Jobs's

choice of disk drive, the piece of computer equipment analogous to a record turntable. This was a particularly touchy issue at Apple, since the Lisa haughtily eschewed any of the standard disk drives, instead introducing an odd alternative of Apple's design that never quite worked. Already the Macintosh people realized that the "Twiggy" disks on the Lisa were fidgety albatrosses. Despite Jobs's wishes, Rod Holt, the analog designer of Macintosh, believed that a new disk drive by Sony was superior. (The analog designer is in charge of hardware other than circuitry and chips—like the disk drive, the picture tube, and the power supply.) He and other like-minded engineers continued the relationship with Sony even when Jobs ordered them to terminate it. The high point in this deception occurred when Jobs dropped into Bandley 3 at the same time a Sony executive had come for a meeting. The Macintosh engineers literally ushered the disoriented Japanese businessman into a closet until the Reality Distortion Field had passed.

Once again, the unmasking of the plot was a welcome escape to a hole Jobs had dug for himself. The Sony drive was a substantial addition to the Macintosh mystique of leading-edge yet compassionately designed technology. Not only did it use smaller disks than the previous standard, but the fragile Mylar that held the data was protected by a rigid plastic coating. For the first time, one did not have to handle floppy disks (people still called them floppies even though that adjective no longer applied) like fine china. It was akin to the difference between reel-to-reel tape and cassettes. In addition, the new floppies belied their size by holding more information. (I recall that the physically larger Apple media—bigger than the head of the average fly swatter—held so little information that when I was writing *Hackers,* a single disk could not hold a forty-page chapter. A single Macintosh disk, on the other hand, could hold ten of those chapters! And fit inside my shirt pocket.)

When Jobs discovered these two perfidies, he did not fire the employees, but had the sense to realize that his minions had bailed him out. And since the credit accrued to him, he held his peace.

Some years after the computer was launched, former Mac team member Steve Capps told me that Bandley 3 was "The best thing I ever did in my life." He was echoing the sentiments of almost every one of his coworkers. Then I asked him about what he gleaned from his experience with his boss. "What did I learn from Steve Jobs?" he repeated. "That ignorance [of what you can't do] is great. We learned to keep on trying and trying. We weren't the best, but we tried the hardest. We were just a bunch of lucky nerds."

The "Preliminary MACINTOSH BUSINESS PLAN, 12 JULY 1981," was the first since Steve Jobs commandeered the project. Apparently produced on an Apple II, it used a crude sort of graphics that the Macintosh would soon render obsolete not only in corporate business plans but in the homework reports of high school students. The plan assumed that Macintosh would ship in 1982 for a retail price of around $1,500 and sell 2,245,000 units between 1982 and 1985—an annual rate of 563,000, or 47,000 each month.

The last page of the plan was a whistling-past-the-graveyard joke: a drawing of Orson Welles, ponderously savoring a glass of wine. The legend above the picture read, "We will announce no Apple before its time." The idea of delaying announcement dates would soon hold no mirth at Bandley 3.

The business plan tread with uncharacteristic tact when it came to the issue of Macintosh's position in the Apple product line, particularly its comparison to Lisa. "Imagine two posters next fall," it propositioned, "the first appearing in retail dealers and Sears. The message: 'Apple II has evolved into two new products, each one the best in its class and both low cost. Buy one . . . Or both!!' And a second poster for Lisa dealers positioning Mac as 'Lisa's younger brother.' "

Nowhere was it written that Macintosh would render Lisa an expensive and rather bulky doorstop. Ostensibly, the Lisa and the Mac people were on the same team. In reality it was quite different. Since the Lisa used so much more memory and disk storage than the Mac, it was impossible to run Lisa software on the smaller machine. (Not that the Mac wizards would have allowed it—they thought Lisa's software was too bloated and inelegant.) Any corporation or individual considering Apple technology would thereby have to choose between the two.

It was a battle that Lisa could not win. Not only was Macintosh slated to cost a fraction of Lisa, but the general mentality of Lisa was hamstrung. The Lisa engineers were stuck in the H-P mindset: a conservative ethos designed to produce dependable, competent technology. This affected even the exceptions to the general demeanor: Bill Atkinson later admitted that he and his more daring colleagues "were afraid of our [corporate] customers—we didn't want to offend them. We erred on the side of sterile." A trivial but telling example of this self-censorship came with Lisa's trash can icon—originally the drawing had a little fly buzzing around the can. This was deemed "too groddy" for the suits. The cumulative effect of this conscientious blandness denied Lisa a distinctive personality, which limited the fervor of its users.

In contrast, the Mac team was off in the ozone, designing a computer that fit their own woolly sensibilities. They felt free to festoon the machine with all sorts of loony filigrees. They even coded little tricks deep into the software, including a hook to evoke a mysterious figure named Mr. Macintosh who could suddenly appear on the screen, wave, and disappear, causing the user to think he or she had seen a mirage. The Mac team's synapses still fired to the cadence of the 1960s; most of them had managed to catch the tail end of that social revolution and were still hungry enough to want more. Skirting the lip of hubris, they believed that their efforts could cause a reprise of that revolution—engineering itself would explode into art. How could the Lisa draftsmen compete with the cubists, the surrealists, the abstract expressionists of Macintosh?

"Basically," Randy Wigginton, a Mac software designer told me of the Lisa group, "we tend to think we're better than them."

Later, Larry Tesler (who was to become an Apple vice president) would claim that accounts of the Lisa-Macintosh rivalry were "exaggerated," explaining that "as in any friendly rivalry, some individuals took the competition too seriously. By and large the teams gave each other both moral and technical support. Half the Macintosh programmers came from the Lisa group, and most of those were working on both Lisa and Macintosh at the same time." But Steve Jobs, undoubtedly one of the overboard individuals to whom Tesler was referring, consciously seemed to use the Lisa division as a punching bag for the Macintosh crew. Jobs made the competition into a direct challenge when he bet the Lisa Division head John Couch five thousand dollars that Mac would ship before Lisa. (Introducing the lower-cost Mac before Lisa would have smothered Lisa at birth.)

By the time of the Mac launch, however, even when Steve Jobs attempted to be diplomatic, he damned the Lisa with faint praise. "Remember, Lisa was the first time," he told me. "I guess I encourage the Mac group to understand they're the best in the world, so they tend to criticize other things, as I do, too, and that's okay. But it's also good to understand that most people [in the Mac team] have been able to stand on the Lisa people's shoulders, maybe avoid some mistakes. The Lisa people wanted to do something great. And the Mac people want to do something *insanely* great. The difference shows."

Steve Jobs lost his bet that the Macintosh would be finished before Lisa. He didn't even come close. Macintosh's course was plagued by setbacks, and Jobs and the Mac confronted constant frustration.

One of the bigger delays came as a result of software director Bud Tribble's departure in November 1981. All along, Tribble had been on hiatus from medical school; now, his advisers told him that if he did not return, he would be drummed from the program. It was a torturous decision, but he left Apple. Andy Hertzfeld recalls being so stunned at the defection that he felt the entire project might fall apart. Finding someone to step in was not easy. Jobs not only needed someone with deep engineering knowledge but the attitude required to go with the spirited flow of the team. He would test applicants with questions like, "At what age did you lose your virginity?" or, "Have you ever taken LSD?" At one memorable interview, Jobs, Smith, and Hertzfeld found themselves across the table from a very straight applicant who was visibly shaken by these queries. Jobs began to gobble like a turkey, and the trio broke up laughing.

The man Jobs chose for Tribble's post was a former PARC engineer named Bob Belleville. He was soft-spoken, contemplative, very smart, and provided a necessary stability to the high-flying team. But his very professionalism—his adulthood—set him apart from the wizards he was charged with managing. Simply put, he did not see the Macintosh project as an expression of rock and roll. The others did. The tenders of the Macintosh soul despised Belleville, particularly Andy Hertzfeld, who ten years later would still launch into uncharacteristically venomous anti-Belleville tirades at the slightest provocation.

For much of 1982, the Macintosh team hoped to launch at the National Computer Conference in March 1983. This seemed reasonable, as the basics of the computer, including the casing, the toolbox, and the design, were essentially frozen. Since the conference would be held in Anaheim, plans were discussed for "tying in a Mickey Mouse/Disneyland theme with Macintosh . . . to create a lasting personality for the product that will transcend the technology and pique the interest of middle America."

Earlier that year, in a more optimistic frame of mind, the Mac team triumphantly had scrawled their names in a design mold. It was Jobs's idea—they were artists signing their work. The signatures were to appear on the inside of every Macintosh computer. No one, except stray repair technicians, would see it. But it meant a lot of them.

Mickey Mouse and middle America, however, were left at the altar—the announcement date was bumped once more. A series of problems forced the team, essentially, to redesign the computer several times. The number of dots on the screen had to be increased, to display full lines of text in fine resolution. Several schemes for disk drives were adopted and discarded.

Frustration mounted with each delay. People complained of "constant time toward completion." Bill Atkinson later compared it to running a nightmarish footrace where, each time you approach the finish line, some unseen force catapults it a huge increment beyond your reach.

Essentially, Burrell Smith wound up designing five different Macintoshes, each one reportedly a tour de force of engineering. "Burrell was the central figure of Mac," recalled Andy Hertzfeld. "His first prototype was the seed crystal that attracted the rest of us, and every version thereafter was built around a core of brilliant hardware." Each iteration of the Mac was better than previous, each one squeezing more performance, and more insane greatitude from the relatively modest hardware.

Through late 1982 and all of 1983, the pressure kept increasing in the mini-Manhattan Project at Bandley 3.

Jobs was careful to keep the team small—he insisted that the software design squad should not exceed ten. That way, there was a guaranteed absence of bozos. One of the best additions turned out to be Bill Atkinson, an unofficial defector from the Lisa team. Bill was definitely rock and roll. He had devoted three and a half years of his life into creating QuickDraw, and the intensely emotional programmer was discouraged at the dawning realization that the pricey Lisa was a commercial flop. In addition, he became embittered at Apple's failure to emphasize in its grinding Lisa media campaign any of his considerable contributions. His consolation was an Apple Fellowship, a rare and mysterious honor that is the company's equivalent to the Congressional Medal of Honor. (A Fellow is given carte blanche to work on projects, as well as lucrative stock options.) But he realized that only with the Macintosh could his work reach a mass audience. Without formally giving notice at Lisa, he set to work on MacPaint, the spectacular example of Macintosh's graphic abilities.

Other veterans from Lisa and Xerox found their way to the team. One was Bruce Horn, a lanky blond programmer who had hung out at PARC as a teenager. Horn was assigned the job of writing the Finder, the program that would improve upon Lisa's Desktop Manager. During that same period in early 1982 Larry Kenyon came over from Lisa to work on operating system tasks. His wife, Patti, was also working on the team. Such were the delays that the Kenyons completed an entire development cycle on a human being—Macintosh's first birth—before the computer's launch. Then there was Steve Capps, whom Jobs had lured to Bandley 3 after seeing a graphically impressive chesslike game Capps had hacked on the Lisa. Capps was a stocky, gregarious upper New York State native who had acclimated totally to California, traipsing around in trademark shorts and high-top sneakers.

In early 1983, Jobs also authorized an unusual addition to the team: Susan Kare, an in-house visual designer whose main responsibility was the "look" of the Macintosh. It was her job to imbue the Macintosh screen with uniformly attractive and functional images. A young woman who had attended the same suburban Philadelphia high school as Andy Hertzfeld, and had migrated west to work for an art museum in San Francisco, Kare had a particular talent for creating icons. Her equivalent of canvas was the little block of pixels 32 by 32—around a thousand dots that would be blacked or left blank. Of course, as dictator of design, Jobs approved every icon. At one point, he rejected Kare's little picture of a rabbit, complaining that it looked "too gay."

Kare eagerly accepted the concept that the Macintosh should have a whimsical side to its personality. "I'm not trying for cutesy, but something a little different," she would say. She knew she was in no way a technical wizard like those in the cubicles surrounding her, but she gently pushed for more latitude to refine the Macintosh as a rounded if somewhat quirky work of art. When the computer came on, the first thing someone would see was a tiny self-portrait of the Mac, with a smiling face to indicate that it had successfully performed a memory scan and all its chips were in order. When users set the alarm in the internal clock, they would click on a picture of a rooster. And when the machine crashed—as it did, too often—a dialog box would appear with a picture of a bomb. (This image actually made some people go berserk with rage—in their view, not only was the computer failing them, but rubbing their faces in it!)

Besides icons, Kare worked on the look of the Macintosh programs themselves. She set about refining details like the look of the title bar, that border on top of a window, giving it distinctive pinstripes. This was far more than a cosmetic makeover. It was partly a careful accumulation of nitpicky details—frills, pinstripes, curlicues, and the gray tint in the scroll bars—that established what has been called the "look and feel" of the Macintosh. Compared to the phosphorescent garbage heap of DOS—an intimidating jumble of letters and commands—the world one entered into when flicking on a Macintosh was a clean, well-lit room, populated by wry objects, yet none so jarring that it threatened one's comforting sense of place. It welcomed your work.

The sedateness and elegance of the Macintosh gestalt could be punctuated by exciting events. The beep when the machine is turned on. The sudden appearance of a drop-down menu. The darkening of an icon when the file or application it represents is not available at that moment. The zooming animation as the windows open and close.

And if any questions arose between you and the computer, lines of communication were flung open wide by a standard-looking dialog box, which took your hand to the next step. You would use the mouse to slide the cursor over the proper "button." If your choice was the one deemed by the designer the most obvious, it would be double-bordered and you would know that by hitting the return key, that choice would be made. (For instance, during the "save file" process, the button labeled SAVE would have a double-thickness border and the CANCEL button a single border; hitting RETURN saved the file.) These dialog boxes would appear in every Macintosh program, for almost anything. This was part of the topology of Macintosh, part of the Macintosh religion. All religions have their look and feel; one glance at a cathedral and you are swept into the gestalt of the Catholic religion. Likewise, a peek at a window framed by title bar and scroll bars is enough to evoke the sacraments of Mac.

Susan Kare's other important task was the look of the fonts. For the first time, typefaces mattered on personal computers. The Macintosh, straight out of the box, was going to offer the user six or seven different type styles, and these were all to have variations in italic, boldface, and even esoteric variations like shadow and outline. To save licensing fees, the fonts themselves were not the copyrighted typefaces offered by the big production houses, but knockoffs that Kare would design—versions that looked like Times, Century, Helvetica, and even Gothic.

There was some controversy at Bandley over what to name these. Kare suggested typefaces based on the stops on the Paoli Local train that passed through the Philadelphia Main Line suburbs she and Andy had been raised in: Ardmore, Merion, Rosemont, and such. Steve Jobs liked the idea of naming fonts after cities, but insisted they be *world-class* cities. The eventual urban appellations reflected the flavor of the individual faces: New York, Geneva, Chicago, and so on. The Old English–style font was called London. Then there was a purposely jumbled font, a mishmash of big and small, thick and thin, serif and sans serif. This was called Ransom, since it looked like a kidnapper's note. That was judged too far-out even for Macintosh, and the name was changed to San Francisco.

Again, the seemingly trivial was to have wide implications. Because of Macintosh, people were going to learn more about fonts and typography using the Macintosh than they had since Gutenberg first got his hands all inky. Fonts, which had previously belonged to the equivalent of guilds—those with access to Linotype machines and art directors and printing presses—now were accessible with a single point-and-click. To echo Bill Atkinson's words, the fine china was now available for everyday use.

The Macintosh Product Introduction Plan, dated August 15, 1982, was a weighty document that attempted to position the Macintosh as a "truly mass market commodity . . . [an] appliance computer." The report provided sketches of some archetypical customers. The cast of characters, illustrated by cartoon portraits, included a thirty-five-year-old suburban office professional who wants to be technologically hip but feels intimidated by computers, a university provost seeking a chance to proselytize the information age, a Rotarian in his forties who vaguely thinks that computers can help run his small business, and a University of Michigan student teetering between an interest in Carolingian empire music and her parents' hope that she become a lawyer. All are ripe for Macintosh.

The report estimated first-year sales between 202,000 and 236,000 units.

As the designers hustled to finish the machine, those charged with the encouragement of a third-party software base traveled the country. At the time the Macintosh was first being developed, the industry devoted to creating and selling software products for personal computers had barely emerged from a preindustrial culture of hobbyists swapping little programs and perhaps, if the program was deemed particularly neat, taking out an ad in *Byte* or *Dr. Dobbs' Journal.* The biggest companies were probably VisiCorp, which sold VisiCalc; and Microsoft, whose specialty was operating systems and computer languages like BASIC, though the computer aspired to become a force in applications such as word processing. On the come was Lotus.

Steve Jobs had been feuding with VisiCorp, which was working on its own, somewhat watered-down, PARC-like system. (Designed to run on souped-up IBM PCs, Visi-On was late and lousy, dead on arrival. The company never recovered.) Microsoft was another story. Its leader, Bill Gates, was the premier mind in the industry, and might have been the only person on the planet whose ambition matched that of Steve Jobs. Jobs himself made a pilgrimage to the Seattle area and wove a mesmerizing verbal tapestry concerning the greatness of the Macintosh and how the future was tied to its advances. As part of his awesome spiel to Gates's minions, he talked about the $20 million factory Apple was building in Fremont, California; he claimed that the building would gobble raw materials of plastic, metal, and silicon, and out the door would emerge Macintoshes. Semimocking Jobs's vision, which they came to accept even more thoroughly than Apple realized, the Microsoft engineers gave the code-name "Sand" to their Macintosh software effort, in honor of the raw silicon the factory would be ingesting.

The Macintosh coincided with Microsoft's vision of the way all computer software would look one day. Former PARC wizard Charles Simonyi was one of Microsoft's leading theorists, and he was totally sold on the vision. Not that Bill Gates, Microsoft's peripatetic chairman, wanted to rush things. At the same time that Microsoft was working with Apple to develop Macintosh applications and beginning its decade-long quest to port its graphical user interface to IBM-standard computers, it was establishing an invincible beachhead in operating systems that used the doomed and hideous command-line system of old. Yet Gates knew Macintosh portended the future and this knowledge (not Jobs's sales pitch) was why he signed up Microsoft.

But Microsoft was not just any software developer. So intense was its effort, and so linked to Apple's that it went beyond third-party status. It was kind of a first-party of the second part. In terms of sheer human effort, Microsoft at times had as many people on its Macintosh software squad as Apple had in all of Cupertino.

Gates and Jobs made an interesting contrast. Both were gaining reputations as the key entrepreneurs of their time. But Gates was technically more capable and also had proven himself as the sole leader of his company. Jobs ran on instinct and charisma, and was kept at arm's length by those who really ran Apple. Sometimes Gates got impatient with him. "The guy should turn it off sometimes!" he said. "He pushes too hard, too much 'us versus them.' When he attacks stuff, he can be rude; he does it almost intuitively, like a religious thing. For instance, he looked at the way we were doing documentation, and starting saying, 'It's a piece of trash, we can't work with you guys . . .' "

Yet Gates ultimately decided that Jobs's role in the Macintosh was heroic. "People concentrate on finding the guy's flaws—why?" he said. "He's in the center of things. They ask—'Does he know the instruction set of the 68000?' I don't think that's super-important. There's no way there would be anything like that Macintosh without Jobs."

Microsoft was only the beginning of Apple's third-party effort. The goal was to have five hundred applications out into the marketplace by the time the Macintosh had been out for a year, and double that the next year. As stated in the product introduction, "we must introduce with two things: a set of software which makes Mac immediately useful in key markets, and the unmistakable impression that Mac is the industry leader upon which all worthwhile software will appear in the future." Macintosh's new marketing manager, Mike Murray, called the people charged with building the third-party development effort "software evangelists." The first was former Hewlett-Packard engineer Mike Boich, followed by Alain Rossmann, then Guy Kawasaki. (All three went on to become entrepreneurs, though Kawasaki somehow wandered astray and became a computer columnist.) Kawasaki later provided the pithiest job description of evangelists: "[We] were Apple's kamikazes who used fervor, zeal, and anything else to convince software developers to create Macintosh products . . . [We] sold the Macintosh dream. The software developers who bought into the Dream (and only some did) created products that changed Macintosh's principal weakness—a lack of software—into its greatest strength—the best collection of software for any personal computer."

The easiest part of their job was taking the prototype to potential software developers. "People were uniformly blown away," Boich told me in 1983. "I remember one guy who had just designed what he thought was a good accounting program—after seeing the Mac, he felt like he had just designed the best propeller in the world, and saw a jet fly by."

Early in the process, the evangelists visited a fledgling Cambridge, Massachusetts, software outfit called Lotus Development Corporation. They were unimpressed with its new product, a spreadsheet called 1-2-3 (though better than its predecessors, it still used obscure commands and codes, stuck in the pre-Macintosh mold of computing), yet Apple's evangelists liked Lotus's founder Mitch Kapor, who greeted them by saying, "I'd sell my mother to get a Mac." Apple later gave him a prototype without that stipulation, and Lotus began working on an ambitious project for the Mac called Jazz. By the time the Mac launched, Lotus had become the world's biggest personal computer software company, having established 1-2-3 as the essential application for the IBM PC. Yet Kapor was crazy about the Macintosh. "The IBM is a machine you can respect," he told me at the time. "The Macintosh is a machine you can love."

. . .

Every so often the frazzled Macintosh team would leave Bandley 3 and check into a hotel for a "retreat." Most often Jobs picked cushy resorts down the coast in Monterey or Carmel. Each retreat began with a lecture *cum* pep talk by Jobs. This set the tone for the event, which would proceed with presentations (on one occasion Burrell Smith announced he didn't have ten minutes' worth of things to say, so he played his guitar to fill out the time), discussion sessions, a guest speaker, and often the sort of rowdy misbehavior associated with heavy metal rock bands. After a display of fires on the beach and midnight skinny-dipping, one resort banished the Mac team from future performances. A second venue blackballed Apple after Mac people commandeered the dessert carts, using the elegant creamy concoctions as fodder for a food fight.

Jobs's speeches were punctuated by slogans. It was on one such retreat that he originally encouraged the Mac team to consider themselves pirates. In another speech he wrote on a large easel that THE JOURNEY IS THE REWARD. Only a cynic would interpret the flip side of this maxim—*no big bonuses.* But there were few cynics on board. The Mac people were consumed in their quest to make history, hyperaware of the specialness of their enterprise. They knew Apple would reap ungodly financial dividends from their labors, but that wasn't the point at all. They believed the Macintosh would be a dividend to the entire world. That is what drove them, made their sacrifice worthwhile, and made the whole thing a jolly little enterprise. "Very few of us were even thirty years old," one of them told *Esquire.* "We all felt as though we had missed the civil rights movement. We had missed Vietnam. What we had was the Macintosh." These were the best days of their lives, and they knew it.

Perhaps the most telling epigram of all was a three-word koan that Jobs scrawled on an easel in January 1983, when the project was months overdue. REAL ARTISTS SHIP. It was an awesome encapsulation of the ground rules in the age of technological expression. The term "starving artist" was now an oxymoron. One's creation, quite simply, did not exist as art if it was not out there, available for consumption, doing well. Was Engelbart an artist? A prima donna—*he didn't ship.* What were the wizards of PARC? Haughty aristocrats—*they didn't ship.* The final step of an artist—the single validating act—was getting his or her work into boxes, at which point the marketing guys take over. Once you get the computers in people's homes, you have penetrated their minds. At that point all the clever design decisions you made, all the twists and turns of the interface, the subtle dance of mode and modeless, the menu bars and trash cans and mouse buttons and everything else inside and outside your creation, becomes part of people's lives, transforms their working habits, permeates their approach to their labors, and ultimately, their lives.

But to do that, to make a difference in the world and a dent in the universe, you had to ship. You had to ship. You had to ship.

Real artists ship.

7

On January 22, during the third quarter of the otherwise unmemorable Super Bowl between the Oakland Raiders and the Washington Redskins, a cut to a commercial turned the nation's television sets over to some extremely weird images. Bald-headed drones marched down a long tubelike corridor to some hellish auditorium. They had shaved heads and pajamalike garments reminiscent of concentration camp victims. Guards in shiny helmets monitored their actions. On a large screen a fierce-looking, almost crazed, speaker—Big Brother himself—was spouting cant that, if you could make it out, was truly frightening:

> Each of you is a single cell in the great body of the State. And today, that great body has purged itself of parasites. We have triumphed over the unprincipled dissemination of facts. The thugs and wreckers have been cast out. *Let each and every cell rejoice!* For today we celebrate the first glorious anniversary of the *Information Purification Directive.*

> We have created, for the first time in all history, *a garden of pure ideology* where each worker may bloom secure from the pests purveying contradictory and confusing truths. Our unification of thought is more powerful a weapon than any fleet or army on earth. *We are one people. With one will. One resolve. One cause.* Our enemies shall talk themselves to death, and we will bury them with their own confusion. We shall prevail!

This is a world clearly dominated by a evil monolithic information beast.

But wait—intercut with the speech we see a solitary figure of hope. A young woman in bright red shorts and a T-shirt with the Macintosh "Picasso" logo. She carries a sledgehammer. At the end of the speech she bursts into the room, races down the aisle, and flings the sledgehammer into the screen.

The Big Brother image evaporates in a white-hot explosion. Apocalypse! The inmates are stunned in slack-jawed amazement. And on the screen—not the screen just destroyed, but the television screens of 43 million people watching the Super Bowl—appeared the following words:

> On January 24th,
> **Apple Computer will introduce**
> **Macintosh.**
> And you'll see why 1984
> won't be like "1984."

This was the notorious "1984" spot. Directed by Ridley Scott, it had all the cyberpunk film noir of his recent cult hit *Blade Runner,* and a more coherent plot—IBM takes a fall. It cost a half million dollars to film. Scott recruited an Olympic athlete to hurl the hammer and London skinheads as audience extras. Apple Computer bought air time for it only twice: once late in December, in an obscure television market somewhere on the Great Plains, so that it would be eligible for the inevitable awards in the new year; and the other during the Super Bowl.

But Apple's board of directors, by and large aging white men unattuned to the semiotics of video clips, had hated it. Apple tried to sell off its commitment. Only at the last minute, when its advertising agency couldn't resell the time slot, did the company finally draw in its gut and give the okay to air the "1984" spot. It turned out to be one of the most famous commercials ever aired—the network news shows even did reports on it. (Later, Apple's John Sculley would speak of the advertisement, as if he had known its power all along, as an example of "event marketing.") Long after people forgot who played in that Super Bowl, they remembered the commercial.

It was Apple's first official public acknowledgment that Macintosh existed.

Real Artists Ship. Those words must have been ringing in the ears of Mac team designers, in Jobs's mocking whiny cadence, for weeks before the scheduled launch. When I visited Bandley 3 in November 1983, the atmosphere was a mix of euphoria and panic. While most key components of Macintosh had been finished for weeks or even months, others were still unresolved, notably those concerning the system software that was contained on floppy disks included with the machine.

The least-finished part of the Macintosh was the Finder, the part of the system software visible to the user. It was through the Finder that the Macintosh magic would first appear, where even the most computer-phobic user could instantly get to work, wielding the mouse to launch programs, put files into folders, delete or copy documents, and, eponymously, find things. Bruce Horn had been working on it for months, and when it became apparent that he would need assistance, Steve Capps began to augment his work. If sending a dozen programmers to Horn's aid would have helped matters, Apple would have done it, but the nature of the Finder was such that even adding a single programmer invited an extra level of complexity. The Finder was a balancing act; a relatively small program had to interact with countless other programs (almost all of which had not yet been written), perform tricky tasks, follow the Macintosh religion chapter and verse, and above all, refrain from crashing. (It's like the soundtrack to films," Horn explained. "If you don't know it's there, it's good.") This was close work. Sending Capps in was like lowering a rescuer into a cave, with prayers that two people would emerge alive.

Since the Finder was so crucial, and so far behind schedule, Capps and Horn did not participate in the orgy of prerelease publicity hoopla. They were de facto exiles, in quarantine across the street from Bandley 3, sequestered in a small room they called the Finder Ghetto. When I visited in November, I could instantly sense the hysteria. The room was littered with junk food detritus. Capps and Horn, however, put on a cheerful demeanor. Horn mentioned some tricky revision they had just implemented. "Here we are, sliding down the guillotine," said Capps, wearing shorts and high-top sneakers, "and we make a radical change."

Not so cheerful was Randy Wigginton, the chief author of MacWrite, the word processor to be shipping with every Macintosh. He looked like death. "Sleep deprivation," he explained, a syndrome that had recently led him to cut his work hours from nineteen hours a day to a mere fourteen. In those rare times that he did sleep, a "monkey" program would be working the Macintosh, typing gibberish (perhaps, eventually, *Hamlet*) into MacWrite. If the program did not crash, that was victory.

Producing MacWrite had been a struggle. The word processor was not only to be one of only two truly practical applications immediately available (the other being Microsoft's spreadsheet Multiplan) but was also meant to stand as an example of how a good program could make use of Mac's interface. This was a considerable burden, especially since producing software for Macintosh was several orders of magnitude harder than for less sophisticated systems. This was something that hundreds of would-be Mac programmers would torturously learn: easy to use, hard to program.

Word processors were particularly tricky. "Displaying a single line of text on other computers is simple—normally you just say, WRITE," explained Wigginton. "But on Macintosh it becomes extremely complex. What font is it? What size? You have to measure each character on the screen. Lines are variable heights. And then you have to worry about handling pictures in the middle of the document."

All sorts of interface issues were still in flux. One of the features of MacWrite—meant to be a standard for all applications—was the ability to save a file under a different name. The open question was, which of the two—the original file or the newly named file—should be the one remaining on the screen, ready for more text? "There are extremely good reasons for each way," Wigginton said, but a poll of the Mac team decided that the renamed file would be the one remaining visible. There were a lot of those questions still hanging.

The state of the Finder, still somewhat amorphous at a perilously late date, made these problems worse. It seemed that every time Capps and Horn made a change, the effect would interfere with something MacWrite was doing, and cause the program to crash.

"We've been working on this for so long that it's a dream [to be almost finished] but in some ways it's a nightmare," said Wigginton. "I'm really terrified. If I really had my way, the word processor wouldn't go out for six months. But it has to be done now. Sometimes at night I wake up in a cold sweat—thousands and thousands of people are going to be using this. To a large extent, I'm responsible for a billion-dollar company."

As the final deadline approached Capps and Horn were staying up fifty-eight hours straight, blasting Dead Kennedys records, gobbling vitamin C like popcorn. Capps would blow off steam by playing video games, furious sessions with Defender, at which he was an adept, second only perhaps to Burrell Smith. Horn would just sit there and scream, top of his lungs.

Working at his home one day, Bill Atkinson became so frustrated at the state of the Finder that he yanked the mouse cord from the machine and threw the pointing device at the wall, making a serious dent in it. "They're going to ruin my project!" he wailed.

Mike Murray liked charts. Macintosh's head of marketing—a Stanford MBA hired by Steve Jobs after minimal experience in Hewlett-Packard's office in Corvallis, Oregon—would get an interviewer into his office, and the interviewer would be embargoed. This meant that he or she could not write a word about Macintosh until January 23 (a day before the official introduction, adjusted for the schedule of the newsweeklies). And then Murray would go to the charts.

He would draw something that looked like a mountain, the Matterhorn perhaps, and identify it as a bell-shaped curve. It represented the acceptance of personal computers. He would identify a point very near the left of the shape, when the slope was still low and proclaim that this was our present location on the curve, with only five percent of the population in that little foothill. These were the risk-oriented, "early adopters." There were only so many of these, however, and everybody else was still distrustful of computers because they were so hard to use.

But now we were on the cusp of the Macintosh era, he'd say, where computers are no longer "an end to themselves, but a means to an end." A tool, an appliance. "A computer for the rest of us." Now, the bulk of people—those who would drive the curve up to the mountain peak, and ring the curve-shaped bell—would be computer customers. These, he would say, echoing Doug Engelbart's prediction of a decade before, were the knowledge workers—the target market for the Mac.

"Our definition of knowledge worker is someone who sits behind a desk and plans to take information and crunch it with ideas," he'd say. Then he would mention the philosophical dream: "The philosophical dream is How can we do something that will improve people's lives?" Murray would pause and confide that, "You don't find that in any other computer company." And then Mike Murray would look at you. He had slightly mussy brown hair and a neatly trimmed mustache (sort of like Paul McCartney in the *Sgt. Pepper* days), and big watery eyes. He was not yet thirty years old and he was in charge of selling a product that would determine the fate of a billion-dollar company. "If we fail," he'd say, "it will be a sad story. I'll cry."

. . .

Steve Jobs was convinced that Macintosh was the final showdown at the Computer Corral. It was Apple versus IBM, to the death.

He was annoyed at having to defend Macintosh from industry pundits' criticism that it was not compatible with software that ran on IBM's disk operating system. In retrospect this complaint seems ludicrous, like criticizing tractors because they do not require oxen. But at that time, with many having declared IBM the unassailable leader in personal computing, the consensus was that launching a computer that did not run IBM-standard software was folly. Jobs, of course, had his own perspective.

"If we don't do it, IBM is going to take over," he said. "If having really great products, much better products than theirs, isn't enough to compete with them, then they'll have the whole thing. They'll have the greatest monopoly of all time. It'll be like owning every oil company and every car company in 1920. If we don't do this, nobody can stop IBM. It's kind of like watching a gladiator going into the arena. It's really being perceived as Apple's do-or-die."

Jobs went on, talking about his motivation. "I'm not doing this for the money," he explained. "I have more money than I can ever, hopefully, even give away in my lifetime. And I'm certainly not doing it for my ego. I'm doing it because I love it. And I'm doing it because I love the people. And I'm doing it because I love the idea of making a great ten-billion-dollar company. If we fail, that will mean my entire worldview is all wrong. And my judgment about people is all wrong. If this is another failure, then I should question my work, I should go write poetry or something, go climb a mountain."

It was clear that Steven Paul Jobs did not expect be composing strophes or scaling K-2. He was already convinced that the battle was over. Apple had won. "Every bone in my body says it's going to be great. And that people are going to realize that and buy it. You see, this job isn't done until we've sold several million of these things."

When do you think that will happen? I wanted to know.

"Eighty-five," he said confidently. "I think we'll break the cumulative two million total in 1985. Sure."

In early January 1984—one week before the absolutely final day for the floppies to ship—the software was still crashing. The frustrated engineers called Steve Jobs, who by then was in New York promoting Macintosh.

Jerome Coonen, one of the software engineers, spoke for the entire team. They had failed. The smart thing was for everyone to get some sleep, come back in a couple of days with a fresh perspective, and finish the software then. Meanwhile, the Macs could be shipped to stores without finished software. In two or three weeks or so, when the rejuvenated team completed a bug-free release, Apple could Federal Express disks to the stores, which would slip them in the boxes. Granted, the machine would not be available for customers to buy right away. But how many Macintoshes were we talking about here? Each of around fifteen hundred dealerships had an initial allotment of two computers. Of those two, the store would probably want to keep one to run demos for future customers, who would buy later in the month when Apple flooded the pipelines with Macs. So, Coonen concluded, all they were really talking about was fifteen hundred possible sales. In a couple of weeks they would get the bugs out, and it would be as if no problem had existed.

Steve Jobs listened, and then he answered: they would ship in one week, period. Whatever state the software was in next Monday would be the final state. They had to fix it.

"We all walked out of the room and said, 'Well, we've got to finish it,' " Steve Capps later recalled.

No one slept for another week. On January 16, when the finished floppies were to be sent out for production, a frantic software team had worked all night, but at 9:00 A.M. the software was still crashing. A few hours later, the bleary-eyed programmers produced a new version of MacWrite that seemed, if not robust, sturdy enough to perform its basic tasks without generated bomb boxes. Good enough to ship.

Within hours, the machines rolled off the lines of the new, $20 million Macintosh factory in Fremont, California, and were set inside the boxes, upon each of which was emblazoned a colorful, Picassoesque line-drawing of the Macintosh. In every box was a main unit, a mouse, a keyboard, software, and a guided-tour audio cassette, where a soothing instructor talked you through the Macintosh basics, accompanied by Windham Hill recording artists, including sonamublent superstars George Winston and Will Akerman. The boxes went all over the country, and were sitting in stores, waiting to be whisked away by customers, on January 25.

They were real artists, at last.

It was not until the very last moment that Apple set the price of the insanely great computer that would determine why 1984 was not going to be like *1984*. For most of the past few months, the Macintosh's cost was the subject of a heated internal struggle. Steve Jobs wanted it $1,995, a price sufficient to make money but low enough to interest masses of users who weren't even thinking of buying a computer; John Sculley hoped to boost short-term profits by pricing it at $2,495. Jobs thought he had won. "I could get $2,495 for every one I [make] for six to nine months," he told me in November. "But I'd rather really make this the next revolution. So we're forsaking some profits." But deep into January, Sculley finally prevailed. Macintosh would cost $2,495.

On January 24, at the annual stockholders' meeting, every one of the 2,571 seats at the Flint Center at De Anza College (a mere microchip's throw away from Bandley 3 and the Apple Campus) was filled, and latecomers had to settle for cyberspace seating, via a specially arranged telecast. The first four rows had long been reserved for the Macintosh team, garbed in the latest of the seemingly limitless variations of the Macintosh T-shirt. At ten o'clock, in strolled Steve Jobs, wearing a double-breasted jacket and a red bow tie. He recited a verse from Bob Dylan's "The Times They Are A-Changin'. " Then, since this was the official stockholders' meeting and only unofficially a high-tech revival meeting, he brought on some Apple directors to conduct some corporate business. The trivialities of a business enjoying its first billion-dollar year disposed of, Jobs reentered and introduced himself. The mounting hysteria in the room as he spoke was reminiscent of Ed Sullivan's crescendoed introduction of the Beatles almost exactly twenty years before.

"It is 1958," he said. "IBM passes up the chance to buy a young fledgling company that has just invented a new technology called xerography. Two years later, Xerox is born, and IBM has been kicking itself ever since."

He continued with his lesson in computer history, citing IBM's initial failure to recognize the market in minicomputers, then personal computers. He mentioned the dire shakeout that was currently decimating the industry, with only Apple and IBM standing as giants.

"It is now 1984," Jobs said, his voice building. "It appears IBM wants it all. . . . IBM wants it all and is aiming its guns on its last obstacle to industry control: Apple. Will Big Blue dominate the entire computer industry? The entire information age? Was George Orwell right?"

"Nooooo!" the crowd screamed back.

At the invocation of Orwell's name, the lights dimmed and the "1984" commercial came on. After that sixty-second apocalypse, Jobs slowed his cadence. He spoke about the amazing attributes of the product he was about to announce. Then the crowd began to come alive again, as the speakers boomed the neo-Wagnerian strains of Vangelis's score from *Chariots of Fire.* And out of a canvas bag, the same sort of bag that Barbara Koalkin dragged into the conference room at Bandley 3 for my own first exposure, Steven Paul Jobs pulled out the computer that meant the future of his company, and portended something for all our futures. "I'd like to let the Macintosh speak for itself," he said.

The synthesized sound capabilities of the machine were up to the challenge. "Hello," it said, "I am Macintosh.

"It sure is great to get out of that bag."

8

It almost failed.

For the first hundred days—the time that Apple considered essential to establish Macintosh as a new standard of computing—sales had run close to the company's projections. Customers willing to live on the edge purchased approximately seventy-two thousand units. But the buyers were mostly charter members of the class of consumer dubbed early adopters. These, as any MBA knows, were well-off technophiles who debated the merits of Saabs and BMWs, who knew what Digital Audio tape was, whose reading habits leaned heavily toward *Scientific American* and Sharper Image catalogs.

Plain old adopters bided their time. Belated adopters kept using their typewriters. The Macintosh factory kept churning out gorgeous boxes that gathered dust in dealerships. The Mac slowly twisted in the wind, and Apple Computer was suddenly an endangered species. "The rest of us" weren't buying.

I recall meeting John Sculley just a day or so before

the hundred-day rollout was completed. I asked whether the Macintosh had reached its goals, whether everything was on track, and he responded, with forced enthusiasm, that things could not be better. But he knew even then that Apple marketing efforts had hit a wall. Before the black period ended, Apple would experience months where it moved no more than five thousand Macs. Less than a tenth of Steve Jobs's predictions.

I recently asked Joanna Hoffman, the first Macintosh marketer, why the computer performed so poorly in its first year. She laughed. "It's a miracle that it sold *anything at all.* This was a computer with a single disk drive, no memory capacity, and almost no applications. People who bought it did so on seduction. It was not a rational buy. It was astonishing that Macintosh sold as many as it did."

As Hoffman indicated, Macintosh in its nascent form had some serious problems. People like myself who bought Macintoshes in 1984 were so delighted with the machine's virtues that we stubbornly tolerated its faults—but late at night, when nobody could hear us, we secretly despaired about them. Science fiction writer Douglas Adams once wrote a neat description of how the siren song of early Macintosh led to frustration:

> I can remember when we first met . . . a group of people [were] crowded around a small beige-coloured box that looked like a toy . . . I watched, at first with mild curiosity, then gradually I began to feel that kind of roaring, tingling, floating sensation which meant I

> had my first experience of MacPaint. But what I (and I think everybody else who bought the machine in the early days) fell in love with was not the machine itself, which was ridiculously slow and underpowered, but a romantic idea of the machine. And that romantic idea had to sustain me through the realities of actually working on the 128K Mac. . . .

Adams was on the mark at identifying the dark side of Macintosh's chipper personality—"ridiculously slow and underpowered." Both faults were a consequence of insufficient resources to store information, both in the machine's internal memory and disk storage components. The combination made the computer very slow—opening an application seemed to take more time than the last minute of an NBA game—and in some ways unwieldy, particularly for a device touted as an ultra-glide journey in previously uncharted realms of user friendliness. At times, it was an embarrassment. The height of indignity was this: this state-of-the-art computer, costing thousands of dollars, put you through a living hell when you attempted the simple act of copying a disk.

Disk-swapping, as Macintosh owners quickly learned, was a new high-tech form of torture. The simple operation of copying the contents of one disk to another was a flirtation with tendonitis. Ironically, as far as the human interface was concerned, this task was easier than it had ever been. (This was not the case with IBM and Microsoft's operating system; with DOS it was extremely easy to wind up with a floppy disk in each hand, asking yourself which was "A:" and which was "B:," knowing that if you guessed wrong, you might wipe out your business plan.) On the Mac, one simply gripped the mouse, selected the icon for the disk to be copied, and dragged the mouse over the icon that represented a new disk. (Like almost every operation performed on the Mac, it was much harder to describe than perform.) The problems came from the hardware inadequacies. With only one disk in operation at a time and too few RAM chips in the computer, the process was akin to using a thimble to transfer water from one bucket to another.

Say you had a few folders full of files on a disk you had named Fred. You wanted to copy the contents to another disk. Call the second disk Fred Redux. Once you had the icons for both disks on the screen, you'd make sure Fred was snugly in the floppy slot and begin the process. First you would wait for Macintosh to read a small bit of information from Fred into its meager memory. Then Macintosh would prompt you to eject Fred and insert Fred Redux. You'd hear the disk whirr as Macintosh accessed the information it had remembered and wrote it on the Fred Redux disk. Then you ejected Fred Redux, put Fred back in, and the process would repeat. How many times would you have to shuttle the two Freds in and out of the floppy slot? A reviewer for *Byte* reported "more than 50 disk swaps and 20 minutes" to copy a single disk.

The problems of speed and storage cast a shadow on

almost every task the new Macintosh users performed. Consider the word processor, MacWrite. Because of the Mac's limitations, it was impossible to construct a file that ran for much longer than eight printed pages. Ever try to write a novel in eight-page chunks? I once sat in on an early meeting of the New York Macintosh User's Group in which several would-be fiction writers had a discussion on how to do just that. Their novels, they agreed, would all be broken into brief chapters. Not an ideal solution.

How did this happen? How could a product which in so many ways was indeed "insanely great" be in other respects simply insane? Hadn't the Mac Team known about these problems?

They had, and they had not. On an intellectual level they had understood what was happening, but their devotion to the Macintosh dream drove them deep into denial. In a sense, they were paralyzed by a shared illusion, generated by overexposure to Steve Jobs's Reality Distortion Field. The problem actually began with Jef Raskin, who envisioned Macintosh as a stripped-down machine, sort of a toaster with a keyboard, with no more memory than the borderline-obsolete computers of 1980. When Steve Jobs ascended to the Macintosh throne, he quickly realized that in order to accommodate Lisa technology, Mac's memory needed bolstering. But he did not bolster it enough.

Jobs's parsimony was rooted in a worldview shaped in

the early days of personal computing, where very little memory meant quite a lot. The first computers discussed at the Homebrew Computer Club, Steve Wozniak's home base, had 4K memory, the rough equivalent of four pages of text. When the Apple II shipped with 48K memory, it was deemed an enormous expanse. Later, Apple II users would buy circuit boards to add 16K more, and then they *really* felt they were humming. So it is not surprising that Jobs felt 128K sufficient.

But there were two errors to his thinking. First, since Macintosh was a bit-mapped machine, considerably more memory was required than with the raster-based Apple II or IBM-PC. When one of these displayed a word, it used a shortcut—there was a single one-byte computer code for each letter, and that chunk of code threw the letter on the screen. But Macintosh (like the Alto that inspired it) did things the hard way. Instead of using a one-byte code, it actually *drew* the letter, pixel by pixel. While this allowed much more flexibility and made it possible to Get What You See, it consumed many more bytes of memory. As a result, much of the Macintosh's memory was devoted to keeping the image on the screen. In addition, the Macintosh system software made its own considerable claims on RAM real estate. Since the interface was constantly on the screen, some of the programs that kept it going had to be stored in the memory at all times. Macintosh RAM was like the national budget—by the time you accounted for the mandatory components like debt, entitlements, and the military, there really wasn't much left. Lisa solved the problem by shipping with over 1,000K memory, a megabyte. The first Macintosh did not solve the problem at all.

The second mistake was overestimating the *cost* of memory. The price of computer memory chips faithfully follows Moore's Law. Named after Gordon Moore, a cofounder of Intel, the law stipulates that in any two-year computing cycle, power doubles and price halves. (This happens because while chip materials are consistently cheap, scientists are constantly figuring out how to squeeze more function on the thumbnail-size silicon wafers.) So the initial costs of shipping a so-called Fat Macintosh with 512K memory might have been steep, but Apple could have reliably anticipated a steep drop in price. As it was, Jobs forbade his hardware team to provide even the capability to expand to 512K. Fortunately, Burrell Smith defied him, and by launch time, Jobs was able to address critics by promising a Fat Mac early in 1985. (In fact, Jobs himself had used a jimmied-up Fat Mac at the official introduction. The original Mac, with its anemic memory limitations, would simply not have been capable of performing its demonstration tasks that day.)

Jobs was also behind Macintosh's lack of a hard disk drive. In early 1982, Joanna Hoffman had written a memo arguing that the Macintosh needed a hard disk drive, more memory, a bigger screen. Later that day, walking down Bandley Drive, she saw Jobs, almost a block away. As soon as he spotted her, he began gesticulating frantically. He yelled that she was a Xerox bigot, enamored of big machines. Didn't she realize that small was beautiful? (Jobs also knew that putting a hard disk in Macintosh meant that the machine would require an internal fan to keep the temperature down, and he thought that the resulting noise would be inelegant.) His arguments brought her around. So compelling was the Reality Distortion Field that a year later, when Hoffman was preparing business plans for the Macintosh, she had on her desk an Apple III computer with a ProFile hard disk (containing over ten times the storage of a Macintosh disk), which was constantly running out of space—and she never considered that this phenomenon suggested a problem with the dogma universally accepted by the Bandley pirates: No one needs a hard disk!

So Macintosh shipped without the proper infrastructure to support a hard disk drive, a device that turned out to be an essential requirement for a Mac. Even Alan Kay, having been lured to Cupertino as an Apple Fellow, wrote a scathing critique of the Macintosh for John Sculley, calling it "Would You Buy A Honda with a One-Gallon Gas Tank?" Sculley, still not quite up to speed about personal computer technology, stuck the memo in a drawer. (Kay had actually intended his screed to be constructive—he called the Mac "the first personal computer worth criticizing," and from Alan Kay this was high praise indeed.)

In retrospect the Mac wizards chalk up the faults of their masterpiece to hubris. "In our efforts to change the world we were a little arrogant and unwilling to listen," Bill Atkinson later explained. "There was a feeling that Lisa [with its expensive memory and hard disk drive] got out of hand; we didn't want fat Lisa-like programs—

too bloated. One of the things we did was set religion, but . . . we never realized how precarious things were—Mac could have died."

Even though my wrists still occasionally ache from those nightmarish bouts of disk-swapping in the spring of 1984, I'm inclined to defend Steve Jobs and the Mac group, and recognize their mistakes and excesses as a consequence of venturing to the edge. If we are willing to concede that the Mac designers were artists, we must allow them the stubbornness with which any paradigm-smashing iconoclast adheres to his or her founding precepts.

Besides, the root of Macintosh's dismal sales performance was not ultimately attributable to its memory and storage problems, but to a fear of the unknown. For many potential buyers, the graphical interface was simply too weird. People often would dismiss the entire gestalt of Macintosh by opining that "I don't really like the mouse," without a fair-minded immersion in the experience. Perhaps the biggest red herring was the Macintosh's conspicuous lack of cursor keys, the four extra keys on the board that move the electronic pen nib right, left, up, and down. Critics would cite this omission as evidence that the Macintosh designers were elitists who had little truck with those in the trenches who actually used computers. These people supposedly cherished cursor keys and took deep offense at their absence on Macintosh.

The critics had a point—cursor keys can be useful—but there are also things to be said for Jobs's intransigence on this point. Donald Norman, in his book *The Psychology of Everyday Things,* discusses various methods by which designers intentionally limit what a user might do. These are called forcing devices. An example would be a keyhole shaped in such a way that one can physically insert the key only right side up.

In Steve Jobs's mind, the proper way to move a cursor was with the mouse. If users accustomed to cursor keys purchased Macintosh computers and maintained their outdated cursor key habits, it would take longer for them to master the mouse, the key component in the interface. And cursor-key jockeys would never come to accept what Steve Jobs believed—that *no* case existed where a cursor key was superior to the mouse. Jobs realized that merely publishing this information in a manual would be insufficient incentive to hew this line. People would succumb to temptation and revert to familiar if inefficient form. So Jobs denied them the opportunity.

The omission also acted as a forcing device for Macintosh software developers. Joanna Hoffman later explained, "We wanted very much to discourage people from porting their current applications over to the Macintosh. We wanted them to develop new applications from scratch, applications that would closely follow our interface guidelines. Cursor keys would be a temptation [to retain old habits]." In later revisions of the Macintosh, cursor keys would not be deemed such an abomination. "When you're trying to spread a religion you have to be pretty strict at first. After you get them converted, you can relax," Joanna Hoffman said, long after the omission of cursor keys contributed to the accumulation of the best software library in the world—based on Macintosh religion.

That software library was slow in coming, however. Creating a new application for the Macintosh was a difficult process. Prospective software developers had to learn the secrets of the toolbox so they could follow the guidelines for human interface. And then they had to make it *look* great—better than anything that existed on personal computers before the Mac. By creating MacPaint and MacWrite, Apple at one swoop had bestowed upon developers both a blessing and a curse—they used the Mac interface so elegantly that any self-respecting developer would be ashamed to release the same ugly, hard-to-use programs that had been the previous standard. (A few did, and were promptly hooted out of existence.) Macintosh finally did get superior software—but it took months longer than anyone expected.

As a result, if one had had any motivation to shoot down the Macintosh in 1984, he or she could not have asked for a bigger target. Forget that its cursor keys were missing in action, and that its adherents—and designers—were a bit, um, flaky . . . *where was the software?* For the longest time you could buy only a paint program, a word processor that couldn't handle a file larger than ten pages, and a rather underpowered albeit easy-to-use spreadsheet. Although Apple sold modems to connect Macintoshes to telephone lines, for months the only telecommunications program available was

MacTEP, written by hacker Dennis Brothers for his personal use and given away free, samizdat-style. ("Maybe I should have put a message in there, 'Please send thirty-five bucks,' " Brothers later joked.) Instead of software, you had Steve Jobs going coast to coast promising that five hundred people were *working* on software.

For all these shortcomings, Apple took its lumps. But what was ultimately the most harmful assessment of the Macintosh came from the people in charge of buying computers for large corporations, the Fortune 1000. These Management Information Services (MIS) managers lived by a unwritten code: one would never, ever go wrong by sticking to whatever IBM called a standard. When Macintosh appeared, they were quick in invoking the code, as if holding a crucifix to ward off a vampire. And then they hit Macintosh with their best shot. This was not, they claimed, a computer.

It was a toy.

Actually, Apple had been terrified that people would dismiss Macintosh as just that . . . a toy. There was some substance to the charge. It *was* fun to use, after all. The wizards designing Macintosh considered it an open invitation to childlike play, and judged that ability among its chief attributes. Early Macintosh marketeers had urged development of a minimum of two "Macintosh quality" games, "unlike the world has ever seen." "The playful side of the Macintosh personality is important for several reasons," read an early business plan. "It further endears the office user to his Mac, titillates the college user, and provides a reason for office types to carry their Macs home to their family." Yet by 1984 the company was ferociously down-pedaling Mac playfulness. Mention games to an Apple publicist and pearls of sweat would appear on her forehead. The Macintosh, she'd stammer, is a serious machine. It was as if the company, by eschewing the machine's perceived childish aspects, was determined to establish the manhood of Macintosh.

But the testosterone issue was lost already. The previous paradigm of computing—command-based, batch-processed, barely coherent—was deeply associated in the MIS community with masculinity. One's virility was associated with the gunmetal boxes and dense, nonintuitive interfaces of those dense beasts. If you weren't familiar with ">A prompts," if you didn't know what CONFIG.SYS meant, you had no hair on your chest. (Even women, when it came to computing, had to have hair on their chests, at least virtually.) And what kind of person used the mouse? A wimp, obviously. Some New Age softie who babbled about using the right side of his or her brain. Columnist John Dvorak contrasted the Mac with the new version of IBM's computer, the AT, and called the latter "a man's computer designed by *men* for men."

An *InfoWorld* pundit in 1984 encapsulated the Mac's problem: "In spite of its impressive capabilities, the Mac simply doesn't have the look and feel of a business computer." The commentator identified a precedent for the Macintosh's dilemma—nineteenth-century steam engines. "Steam-engine manufacturers could have made their machines quieter, but they discovered that potential buyers wanted steam engines to make noise. A quiet steam engine just did not seem like it was working very hard. . . . Those buyers—the technological sophisticates of their day—were responding to subtle, psychological needs. Though they knew, intellectually, that their steam engines were producing power, they needed to know, emotionally—with their senses—that they were getting the horsepower they paid for. . . ." In this analysis, Apple failed by making Macintosh too subtle, too elegant, too easy to use . . . and, in spite of their protests, too much fun. Incoherence, ugliness, and a steep learning curve were indicators that a machine was powerful. Pain meant gain.

Those of us who quickly embraced Macintosh found the attitude depressing. Those in charge of the marketplace regarded computing as a rite of passage, a sort of hazing. Only by acquiring knowledge in this needlessly arcane system could one gain admittance to the society of adepts. It was not a joyous society, but one of stiff-upper-lips.

Macintosh, on the other hand, instantly seemed to attract a joyous cult. Only weeks after Apple shipped the computer, its customers became the machine's key evangelists. Though it is unfair to stick a demographic label on these passionate converts, clearly the most vocal were of yuppie persuasion, embracing the promise of the Mac paradigm with the fervor with which they'd pursued rock and roll in the 1960s. Though the revenues from these proselytizers was more than welcome at Apple, the phenomenon was not entirely welcome in the executive offices at Cupertino. The very existence of the Mac fanatics fueled the resistance of those whom Apple was most trying to woo. Instead of reveling in the spectacle of BMUG, the Berkeley Macintosh Users Group, where four hundred insanely Macintosh people met every week in an auditorium on the UC campus, Apple cringed in horror. They wore T-shirts! What really terrified Apple's directors was the alleged gulf between the people who designed Macintosh—people like Andy Hertzfeld and Burrell Smith and Bill Atkinson, who probably didn't even own suits—and the Brooks Brothers data processing managers who fondly remembered punch cards. Just because the latter were people who just happened to control the purchases of computers in Fortune 1000 corporations.

Poor Apple. It thought that the Macintosh interface—the culmination of the dreams of Bush, Engelbart, and PARC, filtered and improved upon by the canny hackers of Bandley 3—would take the world by storm. When the revolution failed to materialize, its minions were stunned. "I used to think that you couldn't get large numbers of people to really accept personal computers until you had what we delivered—consistent user interface, direct manipulation, modelessness. WYSIWYG, and so on. That you couldn't sell [a brain-dead system like] DOS to a lot of people. Yet DOS outsold Macintosh!" complained Larry Tesler, years later. "What I didn't understand was that most people didn't get to make their own decisions. The mistake we made was assuming that these individuals [in MIS positions] would have in mind the ease of use of the people who would use them. I never believed that they would go by other criteria. I couldn't believe they would spend weeks training people to use a system that they hated [as opposed to the likable, easy-to-learn Mac interface]. Studies show that when people make their own choices, they choose Mac."

Something had to be done. John Sculley knew this, all of Apple knew this. Even the Macintosh team knew this. The good news was that the worst flaws of Macintosh were eminently fixable. Compared to the Herculean efforts required to get the machine out the door, it would be almost a routine task to add memory, the capability for a hard disk drive—and maybe even cursor keys.

Some improvements had been set in motion well before the Mac's launch. From the start, Apple had planned on a second floppy disk drive, to be plugged into a special port at the rear of the Mac—and using it would prevent those twenty-minute disk-swapping-sessions-from-hell. When the drive finally became available in the spring, Mac users were so desperate for it that they muted their complaints about the cost, a steep $400. The second tweak was the Fat Mac, with 512K memory. Apple finally shipped it in September. Those with original Macintoshes had to pay $1,000 to upgrade their computers to the new standard. They hardly had a choice—most of the upcoming software for the computer would simply not run on the original standard.

But more significant work had to be done to truly bring the Macintosh up to speed, especially the rewriting of the internal code so that a hard disk drive could be supported. At that point, however, the Mac team was burned out, in tatters. After the excitement of getting Macintosh out the door had faded, Joanna Hoffman said to me, "Nobody could go back to a job." Those who tried found themselves listless and uninspired. It was postpartum depression, compounded by shock and horror at the computer's failure to conquer the world. "Everyone who worked there identified totally with their work—we all believed we were on a mission from God," said Randy Wigginton, who had worked furiously during 1984 on a revision of MacWrite, then burned out, spending much of the next year sleeping and watching television. "When people didn't buy it, we were majorly depressed."

And the Macintosh, their creation, languished. No one was giving orders to do the obvious fixes, the improvements that would allow their wonderful creation to fulfill its potential. "We bailed out too soon," Steve Capps later admitted. Capps himself, after working with Bruce Horn on a bug-correcting update of the Finder, left Apple and lived in Paris for a spell, racking up $10,000 in long-distance charges with collaborators while working on third-party Macintosh programs, mostly involving music. (One program, Jam Session, was sort of an instrumental karaoke allowing the user to sit in on a musical session simply by pressing keys on the Mac keyboard.)

Bruce Horn had left even earlier; though he would describe the Mac period as fantastic, in the wake of the Mac's introduction to the market he felt unappreciated. "I wasn't happy with the distribution of credit," he said. "And I didn't feel I was fairly compensated. Also, I was ready [to leave.]" He went on to work for other Silicon Valley companies.

In autumn 1984, Burrell Smith and some of the other survivors decided to end the intertia. They escaped to an isolated warren of cubicles dubbed "Turbotown," in honor of a new version of the Macintosh they hoped to design. It would have a more powerful microprocessor, more memory, a very sharp monitor that would use "gray scale" to make images as clear as photographs, and even a hard disk. But inexplicably, the executives terminated Turbo Mac, and that was the last straw for Burrell Smith. He left Apple, and for years was so bitter that he refused to drive his car anywhere near the Cupertino campus. With Mike Boich, the first Macintosh emissary to third-party developers, Smith went on to form Radius, a company making monitors and other Macintosh peripherals.

Even Andy Hertzfeld, who was reputed to bleed the rainbow colors of the Apple logo, left the company. He had never really recovered from his feud with Mac's software director Bob Belleville; it seemed to Andy symbolic of the company's increasingly bureaucratic bent. Now, disgusted with the direction of Apple's corporate regime and unhappy at the fate of the Macintosh, he decided he could better support the computer from outside the company. From his Palo Alto cottage, he began devising a new version of the Finder, one that allowed the user to work on more than one application at a time. But he missed Apple: "For a month after I left, I cried myself to sleep."

And what was Steve Jobs doing? At first, he was busy delivering the Macintosh to various celebrities. He trekked to Mick Jagger's house with Hertzfeld and Atkinson; the Rolling Stone didn't show much interest so Andy and Bill gave a demo to his daughter Jade. Similarly, Jobs gave a Mac to Beatle offspring Sean Lennon for his ninth birthday. An interviewer for *Playboy* tracking Jobs watched as the Apple chairman and the boy slipped away from Yoko's A-List party to play with the computer. Looking over their shoulders were Andy Warhol and Keith Haring. Warhol himself sat down at the machine and moved his hand over the mouse. "My God!" he said. "I drew a circle!" (A year later, Warhol was on stage at Lincoln Center, plugging the graphics of a different computer, Commodore's Amiga.)

Apple Computer's 1984 corporate report featured the putative fruits of eleven of these so-called great imaginations, with stunning black-and-white photographic portraits of recipients and printouts of new work. Kurt Vonnegut used MacWrite to produce the first two pages of his novel *Galápagos* (it is unknown whether the program required him to open another file at that point). Dianne Feinstein and David Rockefeller were less innovative; they or their aides could devise nothing less mundane than schedules and business letters.

After the celebrity obsession passed, Jobs became increasingly bogged down in corporate politics. Not long after the Mac launch, Apple merged the Macintosh and Lisa divisions. This made sense, particularly in light of a redesign of Lisa, which allowed it to run, in a somewhat less than satisfactory matter, Macintosh-based software. (This only extended the life of the doomed Lisa, which would never receive proper credit as the immediate progenitor of the Mac.) Steve Jobs, however, won few friends when he informed the newly disenfranchised Lisa group that he considered most of them to be inferior players on Apple's team.

Now that Jobs was general manager of the company's most important group—a group with disappointing sales, no less—Apple's board of directors could hardly continue to overlook his deficiencies. Thus began a drama that would lead to Jobs's expulsion from the company he cofounded.

Through much of 1984, there were few outward signs of trouble. Sculley and Jobs were a mutual admiration society, gushing to anyone who would listen how much they were learning from each other. A year after the Macintosh launch, they were on stage at the Flint Auditorium again, demonstrating what they called the Macintosh Office. It did feature a wondrous addition to Apple's product line—an expensive "laser printer" that would immeasurably improve the quality of computer-produced documents—but the other pieces, particularly the File Server, which was the heart of the system itself, were simply not there. Not just incomplete, but nonexistent.

Jobs seemed to be living a fantasy wherein Apple's fortunes glowed. Joanna Hoffman, who in late 1984 had been transferred from international to domestic marketing, received a shock when she attended her first high-level meeting with Jobs. She was instructed to prepare a company-wide forecast. But the numbers she was given for Macintosh sales were, she realized, not the actual number of units sold, but the original, hopelessly optimistic, predictions made in early 1984. "Nobody had the guts to tell Steve that his original forecasts were stupid," Hoffman told me.

The real numbers could not be ignored, however, and soon Apple's board of directors began to think that either Steve Jobs or John Sculley—or both—had to go. After a few months of stunned quiescence, Sculley began to fight for his job. He would keep it by neutering Steve Jobs. Jobs, for his part, argued that his leadership was necessary to maintain Apple's status as a visionary institution.

The business press covered the struggle as if it were a combination of Greek tragedy and soap opera. John Sculley was consumed with overthrowing the man most deeply associated with the company, the man who had lured him from his executive position at PepsiCo—seduced him, as Sculley would later concede. Months before, they were the Siamese twins of Cupertino; their grinning pusses graced a late 1985 cover of *Business Week*, and in the background a Hawaiian sunset framed them like Nelson Eddy and Jeanette MacDonald. THE DYNAMIC DUO, read the headline. Now the relationship had descended to vicious memos and clandestine conferences, as Sculley tried to move Jobs somewhere in the company where he couldn't screw things up.

Unlike the clash between Jobs and Jef Raskin, which had implications for the direction of personal computer technology itself, this was no more than a boardroom drama—a power struggle to win the support of a bunch of Apple's directors. The outcome owed everything to the bottom line. For the first time ever, the company took a quarterly loss. Apple was about to lay off twenty percent of its workforce. The board decided finally that Jobs was the problem. Sculley was at fault, too, for not asserting himself, for sticking too close to the orbit of the wunderkind who had plucked him from Pepsi. But the board believed Sculley could right himself and bring serious coin to Cupertino.

On May 31, 1985, as Sculley later told me with steel in his eyes, "I fired him." Jobs retained the title of chairman of the board, and there was some blather about a role as "the champion of Apple's spirit," but in truth he was forcibly exiled from the Macintosh building to some other structure where the phones may as well have been disconnected. After a summer of limbo, Jobs stood up at a board meeting and announced that he was off to start a new company. "It's obvious I've got to do something," he said. "I'm thirty years old."

He called his new computer company Next. It made no dent in the universe.

9

As the Macintosh struggled through 1984, the man who would help save it was quietly traveling around the country with a small stack of the new Sony floppy disks. His name was Paul Brainerd. He was based in Seattle, but he had nothing to do with that area's software giant, Microsoft. Few had even heard of his company, named after an obscure Italian who lived five hundred years before the first Happy Mac showed up on a monitor.

One day that summer, I was hanging out at the West Coast office of *Byte* magazine when Brainerd turned up. He was a gangly man with large glasses, short yet unruly blond hair, and an enthusiasm about software that made him seem more like a techie than a businessman. His disks held an alpha version of PageMaker, a program his newly formed company was developing for the Macintosh. (An "alpha" program is sort of a rough draft of a finished application; not till software is in the subsequent beta stage is it ready for even preliminary evaluation.) Though the program was buggy—it crashed several times during the demo, a common alpha ailment—it was clear that Paul Brainerd's excitement was justified.

Brainerd had once been a newspaper editor, but more recently had been an executive for Atex, the company that made terminals for newspapers and magazines that had "gone computer." He prided himself on being familiar with both editorial and technological aspects of publishing. He saw the field itself at a crossroads. The wave of the future seemed to be high-end machines designed to produce and lay out display ads for newspapers. "You would install eight of these, at around fifty or sixty thousand dollars per workstation," Brainerd later recalled. "They were designed for expert use only—there was a minimum thirty-day training course. They were not very intuitive." Brainerd knew enough about technology to realize something that the bigger companies did not—that all the benefits offered by those expensive workstations soon could be provided by low-cost personal computers. Brainerd began thinking about how a software application could duplicate the work of those deluxe machines—and more. "The concept required getting a number of things right," he later said. "Lowering the cost, yes, but also lowering the barriers that made them hard to use."

In January 1984, coincidentally the month that Apple introduced Macintosh, Brainerd was out of a job. Kodak had bought out Atex, and the Seattle operation was shut down. Brainerd recruited four engineers from the ranks of his former colleagues and started a company. They weren't sure which microcomputer would be the platform for the application to spearhead the new field that Brainerd dubbed "desktop publishing."

In retrospect it seems inevitable that he would find the Macintosh. It was the only personal computer on which the screen display was the same as the final result—What You See Is What You Get. It was also by far the easiest computer to use. What if laying out a publication were as easy as using MacPaint? With Macintosh technology, almost anyone could actually produce a sophisticated document with a small computer.

Yet it was almost haphazard the way Brainerd became linked to the Mac. Driving around the Northwest soon after the Macintosh launch, he and a couple of his company's cofounders decided to make a cold call to Apple's office in Beaverton, Oregon—just to see this supposedly amazing computer for themselves. The local representative promised to visit them in Seattle. He not only kept his promise, but showed up with a Macintosh in the trunk of his car. "Why don't you guys play with it?" he asked. "I don't need it for six weeks or so."

That local rep might have saved Apple. Exposing Brainerd and crew to Mac magic clinched the deal: desktop publishing would be implemented on the Macintosh, and not the IBM PC. Brainerd's program, PageMaker, would be Mac all the way.

It was only after that experience that Paul Brainerd, at the suggestion of publishing technology guru Jonathan Seybold, was introduced to people at Apple who filled him in on an Apple secret—the LaserWriter. Since 1983, Apple had been busily designing an advanced laser printer, much like the sophisticated photocopier-style computer printers that sold for around $30,000. Apple's would cost around $7,000. I admit that when Jobs first explained this to me in late 1983, I was startled—I wasn't sure how many personal computer owners would buy a printer that cost over twice as much as Macintosh itself, no matter how great it was. But Jobs assured me that releasing this printer was going to be one of the smartest things Apple had ever done. (The chairman did not mention that at one point he had adamantly opposed the project.)

Burrell Smith had designed the logic board of the LaserWriter. He approached it as if he were designing a new computer. In terms of its innards, the new printer *was* a computer: a monitor-less Macintosh, only much more powerful than the standard Mac. It's version of the Mac's 68000 Motorola processor made it run fifty percent faster than the Macintosh. It also had three times the memory of a Fat Mac.

But Apple's smartest design decision concerned the manner in which the printer produced its documents. Instead of exclusively accepting the QuickDraw routines utilized by the standard Apple printer, the LaserWriter also made use of something called a page description language, called PostScript, created by software wizards who had recently left Xerox PARC to form a company called Adobe. When a Macintosh program was equipped with PostScript, a visual display of the document would be embedded in a series of instructions; issuing a print command would send those instructions to the printer, which would interpret them to create the document.

This technique was especially powerful in generating typefaces and fonts; PostScript stored these as mathematical formulas. This was not only faster than using bit-maps, but much more flexible—a simple tweak could, with very little computation required, change a style from normal to bold, or switch the point size from three points (very small) to banner size—all without the dreaded "jaggies" that normally characterized output generated from the computer. Instead of using its own knockoff fonts as it did in Macintosh itself, Apple licensed the traditional fonts Times Roman and Helvetica for inclusion in every LaserWriter. (In addition, Adobe marketed entire typeface libraries which could be downloaded into the machine's memory.) PostScript was versatile, and using it, you could generate an array of stunning effects, like words swirling in a spiral.

Best of all, PostScript commands could be read not only by LaserWriters, but other sorts of printers, including professional-quality Linotype machines. So if you had a PostScript file in a program like Paul Brainerd's PageMaker you could send it directly to the big machine that prints out newspapers and magazines. A businessperson's desk in a small office could now be a complete print shop.

All along the designers of Macintosh had said that they could not predict the nature of the revolutions it would ignite—that task would be up to its developers. PageMaker was the evidence—a revolution in publishing.

. . .

PageMaker was amazing. It allowed you to venture into Mac's enclosed world, manipulate columns of texts and pictures to create beautiful page layouts, and emerge from the informationscape with a finished publication in hand.

I once was an editor of a weekly newspaper. I still shudder recalling the recurring nightmare of Tuesdays, when we went to the art house to lay out the paper. Our typewritten stories would have been "spec'd" (given specifications for typesetting) and sent there the day before, along with the pictures we thought we'd use, and, of course, the allegedly photo-ready advertisements our staff had solicited. Upon arrival, we would paste some of the blank "boards" on an angled table, and then begin laying out our tabloid, page by page. People working the Linotype machine would hand us the columns of text, along with the headlines, and we'd figure out what looked good, and begin fixing things down to the boards, using hot wax to hold them down. We'd take a column of text, size it to the page, and snip the excess, to be jumped to another page. It was like a jigsaw puzzle with no set solution except adherence to principles of design (as best we understood them). After a first pass, adjustments began. We'd resize the pictures, sometimes flopping them. We'd fix typos in the copy. We'd rewrite headlines. Someone would call in a change to an ad. Every time an adjustment was necessary, we would give the instructions to the overworked craftspeople at the shop and they would go back to their incredibly expensive machines, returning an hour or so later with the change. This would continue until very late in the day—by then it was usually well into the night—when the accumulated changes threatened to throw the whole system into gridlock. At a certain point, we would stop trying to make the layout better and just do what we could to get the hell out of the place. This liberation usually took place about daybreak on Wednesday.

In the summer of 1984, when I watched Brainerd put the alpha-version PageMaker through its paces, I understood that future editors of weekly newspapers would no longer see the sunrise when they left the art shop. There would be no more art shops. The whole operation had been compressed into the virtual reality of Macintosh.

PageMaker's metaphor was the pasteboard. You laid out your publication by selecting a text file and pasting it on the page. The text flowed into neat columns, with any excess clearly marked, ready to be moved to a jump page. Nuisance tasks became automated. (At our weekly paper, we almost always flubbed one of the page numbers or datelines—but PageMaker remembered these things.) The greatest benefit was the elimination of the penalty previously exacted for corrections or changes. Everything on the virtual pasteboard had the malleability of thought itself, so changes that previously took hours could be performed in milliseconds. Want a headline bigger? A mouse click on a menu item would do it. Resize a picture? Grab the edge of it with your cursor and pull. See a typo? Fix it.

As in VisiCalc, PageMaker's improved efficiency only hinted at its real value. Its true strength was in modeling. The cost of trying new things bordered on zero.

You could virtually publish a thousand editions in cyberspace before you printed a single one. It was even feasible, at the last minute, to rip up the entire layout and try something different. If you suddenly wondered what the page would look like if you blew up the graphic 150 percent and wrapped text around it, you could actually *see* how it looked without squandering an investment of time and money.

No wax on your fingers, either.

Only a few years before, cold type—the switch from hot lead to electronic typesetting and layout on pasteboards—had been posed as both the technological villain and savior in the publishing industry, rendering generations of craftspeople obsolete while cutting costs drastically. Now the entire structure of the cold type process was doomed. It was only a matter of time before professional publishers would switch to desktop publishing. As new refinements to desktop publishing came on-line, it became possible to perform very sophisticated tasks with Macintosh, even color separations and digital editing of high-quality photographs. Glossy magazines and high-circulation newspapers would be "desktop-published." As would this book.

Breakthroughs like PageMaker have two sorts of effects. The first is to increase the ease and reduce the cost of performing previously expensive, time-consuming tasks. The second, and possibly more significant, is to empower people who otherwise could *never afford to do the task in the first place.* With little more trouble and not much more money than it took to mimeograph a newsletter where the text sat on the page like a dead perch, anyone could now produce something with the panache and visual challenge of a professionally produced publication.

Lives were changed. I once interviewed Eliot Cohen, a young man working for the county government in Queens who secretly lusted to become a sportswriter. With a Macintosh and desktop publishing tools, he appointed himself editor, publisher, and art director of a newsletter analyzing trends in New York Mets baseball. He sent it unsolicited to team executives and baseball writers, and even managed to sell a few subscriptions. Showing the newsletter to PR representatives of major league baseball teams, he was able to convince them to grant him press credentials. He soon found himself in the locker room at Shea, interviewing Howard Johnson and Doc Gooden. Writers began to call him for comments, and, eventually, editors of other publications gave him work. He cheerfully shared the credit for his career change to Macintosh.

Hunter Thompson once said, "When the going gets weird, the weird turn pro." Macintosh empowered anyone and everyone with the tools to turn pro.

When seeking a name for their new company, Paul Brainerd and his colleagues wanted to choose something that would not only reflect the content of the product, but its open-ended creative power as well. One day they took some time out from calling on newspaper customers and visited the main library at Oregon State University. The reference librarian pointed Brainerd and his colleagues to some texts on the history of printing.

One of them dealt with a fifteenth-century Italian

scholar and printer named Aldus Manutius. Not exactly a household word. But as Brainerd learned more about this relatively obscure Renaissance figure, he realized that he was the perfect icon for his company. Manutius was born in the small village of Bassiano in 1449 or so; historians don't agree on exactly which year. Very little about Manutius is known for certain until, at around age forty, he abandoned scholarly pursuits in Rome, Ferrara, and Carpi, and went to Venice to practice the nascent publishing trade. He instantly made his mark with a series of books based on the classics. Compared to previous printing efforts, these "Aldine" editions were relatively inexpensive, efficiently designed, and compact, probably the first pocket-sized printed books. His volumes were groundbreaking in the sort of same way Macintosh was—in both cases, the entry point for an exciting new technology was significantly lowered. Manutius is now recognized by the cognoscenti as the father of the modern book. In addition, he is credited with the invention of italic type. This was a significant association, considering that Macintosh would allow millions of people to ditch the underlines they were using in their documents to emphasize words and phrases, replacing them with real *italics,* just like the magazines and books they read.

It was perfect. Brainerd called his company Aldus.

Brainerd, of course, was not the first Macintosh developer to call upon a classical personage for inspiration. Something about Macintosh cried out for lofty comparisons. The technological architects who designed Macintosh consciously intended their creation to initiate, in

John Sculley's words, "a second Renaissance" wherein their marvelous tool would enable creativity to explode, spilling over business documents and electronic doodles like a Grucci fireworks display. So it was no surprise that the original Renaissance would be tapped for ideas. A typical product name for a Mac application: da Vinci.

This same sort of fervor, in fact, characterized the people who purchased Macintosh in the first years. Though Apple was disappointed that only 250,000 or so people had bought Macintosh in 1984, those who did buy were determined to convince the world that they had made a clever, perhaps brilliant decision. Even if others could not see much of a future for Macintosh, they did.

A similar blind enthusiasm persisted in the community of software developers who had been thoroughly "evangelized" by Apple's authorized cheerleaders. Macintosh evangelist Guy Kawasaki came to realize that, with the exception of Microsoft, the key contributions to the still-languishing Mac software base would come not from already established companies but hungry start-ups excited by the possibilities of the friendly Mac interface. Apple was desperate for the Macintosh equivalent to VisiCalc, something so valuable that people would buy the computer solely to run it. It had high expectations for Lotus's product—after all, Lotus 1-2-3 had been the IBM PC's VisiCalc—but Lotus's Jazz turned out to be a dud. Mitch Kapor's charges had clumsily missed the point of Macintosh. In the Lotus view, Mac was a computer for beginners, for electronic dilettantes who still clung to a terror of technology. So

Jazz was the equivalent of a grade school primer, an ensemble of crippled little applications that worked well together, but were only minimally useful. No one bought it. (A little later, Microsoft got it right—it used Macintosh to create Excel—what it considered the most powerful spreadsheet on any computer of any kind. And Excel became an immediate best-seller.) The great new products for Macintosh came from little companies like Living Videotext, which sold a program called Think Tank that let you create dynamic outlines; or a company named Telos Software, which made FileVision, a database that used pictures to organize records.

In 1985, the darkest days of Macintosh, the evangelizing began to pay dividends. Mac developers had spent over a year learning the difficult process of mastering the Mac toolbox and creating a new sort of application, and were excited about what they had wrought. In August of that year, they gathered in Boston for the first iteration of what would become a semiannual pow-wow on insane greatness, the MacWorld Expo. Despite all indications that Mac was doomed—sales were still dismal—the atmosphere was that of a high-tech Mardi Gras, a celebration so steeped in excitement that even the most pessimistic pundits began hedging their bets.

It was a harbinger for a turnaround. Only a month before, on July 15, 1985, Aldus sent out the finished version of PageMaker. The Macintosh finally had its VisiCalc, an application worth buying a computer for.

By that time, Apple was somewhat belatedly throwing all its support to the new application. Originally, Aldus's most helpful contacts at Apple were those involved in marketing the LaserWriter who wanted an example, any example, of an application that would make use of Apple's expensive new toy. But several months before PageMaker shipped, Apple began viewing the concept of desktop publishing in a different light—as something that could bolster the flagging sales of Macintosh itself. Brainerd's liaison at Apple asked him to prepare a marketing report on the concept of desktop publishing. This twenty-page white paper went directly to John Sculley. In the months that followed, Apple paid close attention to Aldus, and when PageMaker shipped, Apple paid some of its marketing costs—everything from cross-country press tours to advertisements in the *Wall Street Journal.* "Apple was desperate to differentiate Macintosh from the IBM PC," Brainerd recalls. "Desktop publishing was their only viable option."

It worked. Brainerd recalls visiting potential corporate buyers who at first would sneer when he pulled the Macintosh out of the case. "That's a toy," they would say. Then he would run PageMaker and it would become clear that the toy was a tool, at least as far as producing visually attractive documents was concerned. To appease the in-house publication people—who by then were salivating with anticipation—the company would reluctantly agree to buy a couple of Macintoshes and LaserWriters. And a strange thing happened. The anti-Macintosh arguments of the Management Information Services dweebs slowly began to lose currency. The superiority of the Macintosh system would win converts.

The workers in the publishing division would discover that spreadsheets and word processors on Macintosh operated with the same intuitive charm as PageMaker, and benefited from the consistency built into Macintosh from the start. Colleagues from marketing, from publicity, from the executive suites—often people who normally loathed computers—would wander over and get a feel for this computer, try out the mouse, and become seduced themselves. They would recommend that the company augment its desktop publishing Macintoshes with Macs devoted to more conventional applications.

"You would see the pattern," recalled Brainerd. "A large corporation would buy PageMaker and a couple of Macs to do the company newsletter. The next year you'd come back and there would be thirty Macintoshes. The year after that, three hundred."

Though this process was initially more a deus ex machina than a conscious strategy on Apple's part, the company belatedly arrived on a name for the process: the Trojan niche. It was as good a description as any for the way desktop publishing saved Macintosh's skin.

Not long after PageMaker shipped, Apple began overcoming its engineering inertia. By January 1986, the Macintosh Plus was ready. At first glance it was much like the original, but the most severe criticisms of the original were successfully addressed. At last, there was sufficient memory—a megabyte. There was a new drive for the floppy disks that allowed each disk to store twice the information. And the innards of the machine had changed too. The tired remnants of the Mac Team, including Brian Howard and Larry Kenyon, had beefed up Andy Hertzfeld's ROM so that the Macintosh Plus could smoothly accommodate the hard disk drives that would be plugged into a special high-speed port on the back of the box. Apple offered its own model for sale, but dozens of third parties very quickly had versions of their drives.

The Mac Plus even had a new keyboard. With cursor keys.

Though bigger and better improvements were in the works, the Plus was a major milestone. While the original Macintosh was a visionary computer, you had to be a visionary to appreciate it. In an effort to create what he called a "crankless computer," Steve Jobs had eliminated the crank of a steep learning curve, but replaced it with a new crank—insufficient memory and storage. Now that crank was gone. Only when the configuration of a Mac Plus and hard disk drive became standard could the Macintosh switch on the ignition and fly.

John Sculley was now in charge of Apple, which essentially meant that he was in charge of Macintosh. Though the Apple II family still brought in more revenues than the newer machine, Sculley and everyone else knew that Apple's original computer, introduced eleven years before, was an old dog, well into its fade toward oblivion. This was clear despite the fanfare given the latest slick incarnation of the machine, the IIc, introduced by a blowout event with banners proclaiming

"Apple II Forever." If Apple was to have any future at all, it had to keep improving Macintosh, so that impressive new applications like PageMaker and Excel would be able to fulfill their potential. All along, the goal for Macintosh had been to establish the computer as an industry standard alongside DOS, the inferior but widely accepted IBM and Microsoft system. It still wasn't clear that this was possible—perhaps the DOS lead was insurmountable.

There were two drastic alternatives available to Sculley; he resisted both. One was to license Macintosh's operating system to any takers. The experience of IBM showed that when competitors, especially Japanese and Korean hardware companies, could produce "clones" of your computers, most of your customers would abandon you for the cheaper goods. No, the system software—the soul of the Mac—was Apple's crown jewel, and would remain closely held.

A less onerous approach would have been to drop the price of Macintosh, making less money per unit but selling many more computers and increasing the Mac's market share. One need only visit a computer dealership and watch a first-time buyer to understand the virtues of this strategy. (It is not clear that Sculley ever did this.) On one side of the store were several different models of the toady machines that ran MS-DOS, including very low-price "clones" of the IBM PC. On the other side was Macintosh . . . gorgeous, a bit exotic, and clearly superior to the competition. The entry-level Macintosh, however, cost almost twice as much. It was not obvious to novice buyers—or even experienced ones—that the higher initial cost would be amortized by hundreds of hours of more pleasurable and effective computing. So Apple lost millions of customers, and forever lost its chance to gain a significant market share. Sculley didn't price Macintosh competitively until 1990, and by then Apple was destined to be forever the "other" computer, with no more than a significant but small plurality of users. It was a hesitation that would ultimately cost Sculley his job.

Still, in the formative years of Macintosh, Sculley was actively remaking Apple into a large and successful company. At every turn, he attempted to distinguish his leadership from the swashbuckling legacy of Steve Jobs. While he celebrated the achievement of the Mac team, he disdained the pirate ethic that drove them. "The heroic style—the lone cowboy on horseback—is not the figure we worship anymore at Apple . . ." he wrote in his autobiography. "Originally, heroes at Apple were the hackers and engineers who created the products. Now, more teams are heroes."

This was not exactly true. A close look at the manner in which the key products evolved at Apple in the Sculley era shows an interesting pattern: time and again, the crucial advances were made by individuals or pairs of wizards who passionately followed their instincts, often pursuing their goals even after their managers had canceled the projects.

This was partially a result of the management style of the man charged with producing the next generation of

Macintosh technology: Jean-Louis Gassée. He was the first and by far the most prominent of many of the protégés John Sculley would pluck from the ranks of Apple executives (most often from Apple outposts abroad), and occasionally from other, duller companies. Sculley would latch on to these golden boys with the fickle intensity that studio executives bestowed upon a Hollywood flavor of the month. For a year, perhaps even two, this or that executive would be touted as a dynamic presence on the Apple team, and then one of Sculley's fabled and frequent corporate reorganizations—"reorgs"—would displace the stunned executive.

Gassée was for a long time the exception, the only one with a personality that even began to address the charisma gap opened by Jobs's departure. He was a Frenchman, darkly handsome and dripping with Gallic self-assurance, a *noir* visionary in leather, a businessman who seemed to fancy himself a combination of Belmondo, Sartre, and Tom Watson. As head of Apple's French operations, he had outsold IBM, in the process shaping his own persona as a generic symbol for the future. His face turned up in *Vogue,* and he made commercials for mineral water. He had even written a book called *The Third Apple,* which was not merely a book, he wrote, but "an invitation to voyage into a region of the mind where technology and poetry exist side by side, feeding each other." One of the phrases that technology fed to poetry (or was it the other way around?) was that the Apple II "smelled like infinity."

But Gassée was no mere blowhard. He understood

what it meant to be an emissary of Macintosh: you had to channel magic. He knew that in the wake of Steve Jobs's traumatic departure, a sense of play was missing in Cupertino. He believed that one reason Apple was suffering in 1985 was that it had somehow abandoned its roots. Though he loved the Macintosh, he felt it lacked some of the features that had made the Apple II the leader of its day.

These included color and slots. Certainly, color display monitors had always been planned for Macintosh, but this capability had been shelved because it would make the original Mac too expensive. Slots were more controversial. A "slot" is actually a socket on the main logic board inside the computer. If a computer has slots, wireheads describe it as "open," that is, the computational core of the machine (the "bus") can be accessed through those slots. You do this by plugging in special circuit boards. These boards have various functions: some enable you to network with other machines, some enhance video quality, some add memory, some add processing speed, others may include the equivalent of a whole new computer, allowing, for instance, a Macintosh to behave like an IBM PC so you could run software for either machine. The Apple II was famous for its slots—people cracked the case open all the time and stuck a variety of weird cards into the six slots; they loved the intimate contact the process afforded with the exotic-looking chips and resistors in the Apple's viscera.

But the original Macintosh, by Steve Jobs's fiat, was slotless. Jobs thought slots were inelegant, a remnant from the Homebrew hobbyist days where the first instinct of the mad-scientist–hacker types there would be to jack up the hood and critique the bus. There was also a practical reason for disdaining slots—system crashes. It's much easier to write robust, or stable, applications programs if you know that everyone's computer is configured the same way. Once that consistency is lost, every time a program runs it has to be instructed thoroughly about the configuration on this particular computer, at this particular moment. It's a mess, and this is partially a reason why both the Apple II and the IBM PC were temperamental machines. Jobs was convinced that the ports in the back of the Macintosh were sufficient to handle any peripheral that Macintosh users would need. After all, the Mac was for knowledge workers, not wonks. For that reason, Apple broke tradition and did not include a programming language along with the machine. Macintosh was for drivers, not repairmen. In fact, Mac owners were sternly informed that only authorized dealers should attempt to open the case. Those flouting this ban were threatened with a potentially lethal electric shock. (Talk about a forcing device!)

Gassée reversed all this. He wanted a Macintosh with color, big monitors, and slots. He wanted a powerful machine that would impress even the macho MIS workers who were still snubbing the Mac. Gassée himself had more than a streak of this same sort of machismo. He made no bones about his desire for power, for openness, for slots—he considered it one of his features, not a bug. He presented the prospect to Apple's customers as a solemn vow. The license plate on his Datsun 280Z read OPEN MAC. The only problem was delivering the goods.

Here is where Sculley's team concept broke down. Sculley envisioned the new Apple as a cluster of team players, each squadron consisting not only of engineers but marketers, human resources people, and bean counters. But the road to the new Macintosh, or the Macintosh II as it would be called, was less a cooperative effort and more like a road rally on Baja, a long, rutted unmarked trailway dotted with broken-down four-by-fours. Ever since the original Mac had shipped, various Apple hardware artists had been trying to hatch a supercharged sequel, but somehow, with the boardroom politics, Sculley's frequent re-orgs, and the steady exodus of disenchanted wizards, their projects languished. There was Burrell Smith's Turbo Mac, and then Burrell was gone. There was the Big Mac of Rich Page (the designer of Lisa), and then Steve Jobs cajoled Page into being one of the Next team. There was another mysterious Macintosh called Jonathan, but that also somehow found itself by the wayside.

And then there was the dark horse in the field. It was called Little Big Mac.

Its auteur was a wispy, bearded engineer in his twenties named Mike Dhuey. He had most recently been associated with Apple's File Server, an essential component of the so-called Macintosh Office announced in January 1985. The product was never completed. Its failure was a shorthand for the fortunes of Apple in early 1985. Dhuey believed that out of the File Server's ashes should come Macintosh II, or as he called it, Little Big Mac: an expandable Macintosh. The proposal came solely from his own initiative; he simply wanted to be the person who designed Apple's next computer, and he began doing it. He hooked up with a hardware designer named Brian Berkeley whose thinking was parallel to his, and the two of them embarked on what they called "an underground thing." Though Dhuey drove a Porsche and Berkeley was a Mercedes man, the pair got along well, swapping design notes and stuffing boards. Fearful of Steve Jobs's loathing of slots, however, they kept things quiet. In their memos they never used "the s-word."

Apple's products czar Jean-Louis Gassée came across the project and benignly permitted it to continue, albeit as sort of a "background skunkworks," as Dhuey later put it. Looking at a photo in Dhuey's cubicle of the engineer's hometown skyline, the Frenchman renamed the project "Milwaukee." Subsequent code names of the Macintosh II included "Reno" (in honor of the slots), the overly militaristic "Uzi," and, ultimately, "Paris," an homage to Gassée. For the next few months the designers tried to figure out which features to put on their computer, hoping that it would win out over the other potential Open Macs going at Apple. There were false starts. After some vacillation, Dhuey decided that the Mac's original microprocessor was underpowered for a color machine—the next generation, Motorola's 68020, would be required. Until very late in the process, Berkeley dictated that the power supply would be placed in the monitor—a surprisingly wrongheaded solution that would have limited the range of screen options available to Mac II users.

Gassée finally made his decision on which of the digital contestants would become Mac's big brother. "It was a question of people," Gassée said. "I felt that Mike Dhuey was capable of doing it." Instantly, the other projects were back-burnered or axed, and a phalanx of analog designers, software wizards, hardware nerds, ROM ninjas, and marketers were on the case. A "champion" was anointed to manage the project. T-shirts were printed. Parties thrown. Deadlines set, and missed. Nothing approaching the original Macintosh's Constant Time to Completion, but still worrisome. November 1986 became January 1987, and finally, March. By then it was clear that the Macintosh II would be the biggest thing in Macintosh since the Macintosh.

Not that the two could be compared. The original Macintosh was the anomaly, created by those eager to roil the surface of history. That was its strength. Its failing came in the inevitable shortfall between reach and grasp. Under Sculley and Gassée, Apple concentrated on the grasp part of the equation: having produced what its artists wanted, the job from then on was to satisfy customers. The original artists of Bandley 3 would sniff derisively at each iteration of the technology they had wrought, but ultimately came to admit that the impact from their labors would only be realized when more people used it.

"Steve Jobs [when working on the original Mac] thought that he was right and didn't care what the market wanted—it's like he thought everyone wanted to buy a size nine shoe," said Dhuey in 1987, when his creation was officially released. "The Mac II is specifically a market-driven machine, rather than what we wanted for ourselves. My job as an engineer is take all the market needs and make the best computer. It's sort of like musicians—if they make music only to satisfy their own needs, they might lose their audience."

At the same time Apple introduced the Mac II, its first "modular Mac" with separate monitor, it also unveiled the SE, a variation on the distinctive original-shape "compact Mac" family. The SE was similar to a Mac Plus, but with cosmetic variations such as a color change (from beige to a lighter shade called "platinum"), a slightly faster processor, and, the main wrinkle, a single slot. (Despite the latter, users of compact Macs were still warned to allow only authorized dealers to open the case.) Over the next few years, Apple would introduce a steady stream of new Macintosh models. With each one, Apple strengthened its contention that the mission of Macintosh was finally on course. Yet the press conferences introducing these models were often testy affairs. A continuing bone of contention was Apple's insistence of charging top dollar for its computers. Few would let Apple forget that it had introduced Macintosh as "the computer for the rest of us," and wound up selling computers that often cost fifty percent more than its competitors.

In 1989, for instance, when Jean-Louis Gassée unveiled the SE/30, a compact Mac with a more powerful processing chip at its heart, the Frenchman parried the journalistic thrusts with an edge of hostility. "We don't want to castrate our computers to make them inexpensive," he responded icily to one question about price. "We make Hondas, we don't make Yugos."

Fortunately for Gassée—and Apple—yet another isolated team of heroes with hopes of designing the next Macintosh had been forging on despite the wishes of management. This was a duo consisting of H. L. Cheung, an engineer who had originally been hired in Singapore; and Paul Baker, who had once worked on the Lisa, left Apple to work for Jef Raskin's ill-fated Information Appliance company, and returned. Sculley and Gassée had just canceled Cheung and Baker's project, which was built around the vision of making high-quality Macintoshes for a much lower cost—but the engineers were loath to abandon the idea. "The key to getting something accepted [at Apple] is to make it a *fait accompli,*" Baker said. Only when the pressure from outsiders for lower-cost computers became overwhelming did Apple executives finally discover that such a machine was already well into development. As happened with both the original Macintosh and the Mac II, the executives moved the once-obscure project onto the fast track, and before 1990 was out Apple finally had a series of low-cost Macintosh computers, including not only the Cheung-Baker color LC but a compact Mac—the Classic—built on similar principles. The Classic had all the power of the SE, but retailed for under $1,000—the price point envisioned by Jef Raskin for his Swiss army knife computer.

The appearance of these economy models was one

more boost to the momentum already generated for Macintosh. After taking almost five years to sell a million Macs, Apple would sell around ten times that in the next half-decade.

It had taken until the 1990s, but Apple's lofty goal was finally reached—Macintosh was indisputably the third great standard in personal computerdom, after the Apple II and DOS. And the only one that anyone but a wirehead could love.

10

"If I weren't in business," John Sculley once wrote, "I'd probably be an artist."

Actually, Sculley believed he *was* an artist. An artist, a scientist, and a visionary. Apple's chairman in the post-Jobs era, born in 1940, had studied architecture at Brown University. But instead of designing buildings, he pursued an M.B.A. at Wharton. After a brief tour at the McCann-Erickson advertising agency, he went to work for PepsiCo—he had married the chairman's daughter—and determinedly made his mark on the company, working on the Pepsi Generation campaign, reorganizing the snack foods division, masterminding the Pepsi Challenge. Later, he would speak about these tasks as if they were Great Works, but he must have known better. When Steve Jobs taunted him "Do you want to sell sugared water for the rest of your life?" it marked the turning point for Sculley. Sculley's move to Silicon Valley in 1983—a journey the particulars of which he later accounted with Homeric gravity—was the opportunity to make his own dent in the universe.

In the Silicon Valley and at Apple in particular, bottom-liners in suits were regarded by the true hackers as necessary evils, but irrelevant in the big scheme. Bozos. Sculley, who hated neckties anyway, resisted this characterization. Though he certainly demanded his corporate due—at a base salary of $2 million, and millions more in stock options, his compensation boosted him high on the *Business Week* chart of most-highly-paid-executives—he also sought recognition as a deep thinker, a scientist, the sort of guy who would vacation in Paris to hang out at the Louvre, then, sketchbooks in hand, wander to the Left Bank. Sculley took pains to cite his own thwarted career as an inventor; by his own account, he had, at age fourteen, invented a color television cathode-ray tube. But only two weeks before he filed his patent application, another inventor registered something similar, and this preemptive entry eventually wound up as the heart of Sony Trinitron technology.

Sculley understood that he was particularly well placed to assume the mantle of visionary, since his company was making the most visionary product of its time: Macintosh. Though he did not really take easily to computer technology—close observers would note that when he used a new Macintosh application his hand would literally shake with tension as he grasped the mouse—he did understand that its interface and friendliness were a quantum jump in personal computing. "To John, the Mac interface was sort of the computer equivalent to the pull-tab on a Pepsi can," said Joe Hutsko, who for a time was Sculley's personal technology tutor.

Sculley knew that users hadn't begun to tap the Mac's potential. Desktop publishing was only the opening salvo of a fusillade of developments that would change the way people worked. But through 1985 and 1986 his thoughts about technocultural revolutions had to wait until Apple itself was stabilized. As he told an audience of industry luminaries in 1986, "The very best companies in the world are best not only because of their creativity, but because of their ability to implement. Apple, in growing up, has had to realize that this is an important priority for successful innovation, as important as the creative development of ideas." Compared to Apple's previous ethic—where nothing was prized more than wizardry—this was heresy. But it had to be said. The oft-repeated joke in Cupertino was, "What's the difference between Apple and Boy Scouts? The Boy Scouts have adult supervision." Sculley assumed the role of scoutmaster. After the adults took charge, there would be plenty of time for creativity, for innovation, for vision. And Sculley was determined to be known as the architect of that vision.

So, in late 1985, when Bill Atkinson went to John Sculley with an idea that dramatically enhanced the Macintosh's ability to handle information, the chairman and chief executive officer was a willing audience.

The past few months had been dreary for Bill Atkinson. At first, the inventor of QuickDraw and hero of MacPaint thought he had avoided the post-Macintosh depression paralyzing many of his peers. Capitalizing on the freedom according him as an Apple Fellow, he embarked on a project as potentially transforming as those previous achievements. He called it Magic Slate.

As Atkinson later explained to me, this was the latest iteration of Alan Kay's dream, the Dynabook. It was to be a flat panel of liquid crystal display, about the size and heft of a few legal pads: perhaps an inch thick, and weighing no more than a pound. But it would store enough information to equal an eighty-foot-tower of notebooks. The device would be manufactured cheaply, and sold at a very low cost. How low? "Low enough that we figure you could lose six of them in a year and not be hurting," Atkinson said.

Magic Slate's paradigm was the page. Each screenful represented a page of information. You'd turn these virtual pages by swiping at the bottom of the screen, the same motion you use to flip a page in a notebook. There would be no mouse or keyboard—a touch-sensitive screen itself would be your control device. You would write with a stylus, and your scrawl would instantly be recognized and translated into a clean screen font, ready for editing or digital storage: one more chunk of the datasphere.

When Atkinson told me this, I stopped writing in my own no-tech notebook—a steno pad purchased earlier in the day—and showed him my own scrawl, a script so illegible that only hours after taking notes I am often stymied in deciphering the words. "Could it read this?" I asked. Without an instant's pause Atkinson nodded vigorously.

"I wanted Magic Slate so bad I could taste it," Atkinson said. But as the case with the Dynabook, the idea was too far ahead of reality. He believed that Magic Slate would be possible if Apple had been inclined to pursue long-range research that could develop the requisite new technologies. Apple, though, was more concerned then about fixing Macintosh and boosting its short-term revenues than in pioneering a new concept of product, especially one that even by optimistic estimates would take years to bring to fruition. Magic Slate died.

Atkinson was devastated. While he considered it satisfying to concoct new technology, the exercise seemed pointless if his ideas were not realized and offered for sale in the digital bazaar. "I bought Steve's dream," he said, "making a dent in the universe." For months, Atkinson immersed himself in a numbing void. He spent his days at home, too disheartened to even turn on his computer.

One night, he stumbled out of his house in the Los Gatos hills, wandering aimlessly. The night was cloudless, gorgeous. He spent hours gazing into the sky. Suddenly his own troubles fell into perspective. Who was he? A stray pixel in the eternal bit map. There was no way that human beings were alone in this universe, he thought, and no way that we're the pinnacle of all possible life-forms. The thought was oddly encouraging. Given this fact, it seemed to him, the important thing was what one *could* do. Could you aid your fellow humans, enhance their abilities? Make a contribution?

Bill Atkinson realized that, more than most, he

could. He was in a position of leadership. He was a world-class thinker, a right-brained adept. And, thanks to his feats at Apple, he had the ear of John Sculley.

Following this epiphany, his creative floodgates flew open, and soon he had resurrected some of the concepts of Magic Slate into software that could be quickly realized—on a Macintosh. This new program exchanged the page paradigm for one based on cards, an endless stack of 3 × 5's. The information on these cards would not be limited to text; they could store anything that could be held in digital format, which meant nearly anything: text, graphics, sound, even video. But the most important part of the program would be the manner in which the various cards of information were grafted to each other. At the merest suggestion of a sensible juxtaposition, a user of Atkinson's program could "link" any card to any other. A series of links would result in an information pathway that hearkened back to the dreams of Vannevar Bush, whose influential gedanken experiment, memex, was characterized by the data "trails" that would be cleared by the scientists and researchers using it.

As it turns out, Atkinson was not the only one at work on realizing Bush's vision. The memex vision, of course, had originally ignited Douglas Engelbart, who in turn triggered the series of innovations that would lead to Macintosh. But the most vocal proponent of Bush's ideas was Ted Nelson. For years Nelson had been a peripatic if somewhat cranky figure in the clubby personal computer underground; he was known mainly for the iconoclastic populism in his self-published 1974 book *Computer Lib/Dream Machines,* in which he made a strident and at times hilarious case for truly personal computers.

Nelson had long been an advocate of an electronic publishing industry. "We ought to be able to read and write on computer screens, with vast libraries easily, instantly and clearly available to us," he declared in an underappreciated book called *Literary Machines,* long before it was fashionable to express such things. The key to accessing this information, said Nelson, was the use of what he called *hypertext,* or "non-sequential writing." Nelson's dream for implementing this was (and remains) Xanadu, a complicated system in perpetual development. Vannevar Bush would have approved of Xanadu's key idea—treating the accumulated knowledge of the world as a single body, and providing access to it via a series of dynamic links, under the control of the individual. "A literature is a system of interconnected writings," Nelson wrote, held together by a "flux of invisible thread and rubber bands." While certain experimental novelists (Julio Cortázar with *Hopscotch,* Laurence Sterne with *Tristram Shandy*) had toyed with this notion, Nelson insisted that leaping from document to document was the essential way to partake of the world's ideas. He even outlined a scheme by which copyright fees could be collected by creators of the brief info-gems whose words found themselves on the linked path to enlightenment.

Such ideas were furiously percolating in Silicon Valley in the mid-1980s but it took Atkinson, writing a program for Macintosh, to give them form and substance. He thought his program, which he called Wildcard, could be the ultimate organizer. As often happens in the computer industry, the proposed product name had been claimed by a previous company, and the program came to be known as HyperCard. (After observing the trademark battles in the computer industry for some years, I now envision the English language as a lexical Oklahoma Territory, each word waiting for a homesteader to stake a claim on it.)

HyperCard was meant to occupy a central place in the world of Macintosh, a node from which would tumble all the weird facts, semiremembered appointments, and quickly jotted phone numbers that all too often had previously gotten lost in the shuffle of information clutter. It would be a virtual cockpit; by pushing its buttons and tweaking its instruments one could retrieve the aggregate wisdom of cyberspace—the equivalent of the libraries of Alexandria, all accessible from a familiar "home" card that automatically appeared when you opened the program.

As was the Macintosh itself, HyperCard was a tool that could affect the way you viewed information. By its quick and painless linking process, one could easily see how all sorts of disparate forms of information *could* be linked to each other. Atkinson had a test program with a number of scanned and hand-drawn images; several featured people wearing hats. If you pressed the cursor on a hat, you would find other images on other cards appearing, linked by the visual concept of hats. Lurking behind that activity was an idea—that information could easily migrate across the boundaries of text, graphics, video, and sound. HyperCard was a primer for the digital age.

So if I wanted to make a HyperCard "stack" of personal information, I might have a "home card" reading, ME. On it would be an array of buttons, ranging from ones labeled TIMELINE, PEOPLE, WRITINGS, SCHEDULE, and other paths to branch from. Hitting the TIMELINE button would bring a scroll bar from birth to the present; stopping at a given point would yield a year, say 1988. A list of events from that year would pop up—Publication of Second Book, for instance. Selecting that choice would show a cover of the book, and a row of buttons, giving the choice of displaying a list of reviews, the complete text of the book, sales figures for the book, or a video and audio clips from my promotion of the book. Click on the REVIEWS button and you are linked to a copy of the review laid out graphically as it first appeared; click on the name of the reviewer and you see his or her biographical information. There might even be a RESPONSE button that would yield any letter I wrote commenting upon the review.

As Vannevar Bush had hoped for his memex machine, HyperCard was even capable of charting the sorts of connections an individual mentally mapped out all the time. Your own personal web of links could be seen as a fingerprint of your interests. HyperCard, in effect, taught you about your own brain, the leaps of logic and inference it took. No wonder that when the product was finally released one of its several mottoes was "Freedom to Associate."

Within a few months after his celestial revelation, Atkinson had a mock-up of the program. But he was reluctant to show it to anyone at Apple. He doubted whether the company would pledge sufficient commitment to fulfill the promise of HyperCard. In addition to the psychic pain of the Magic Slate rejection, Atkinson was smarting from Apple's haphazard support of MacPaint. Though Apple had serially assigned several programmers to update the heralded product, the revisions were late and not particularly innovative. Worse, Apple had stopped bundling MacPaint with every Macintosh, depriving Atkinson of the widest possible audience. Atkinson believed that in order to avoid a similar disappointment with his new creation, he would have to leave the company.

No less a luminary than Alan Kay heard about Atkinson's imminent departure. Kay, who had become a sort of high-tech Rasputin to Apple's chairman, alerted Sculley, who immediately requested an audience with the disgruntled software artist. Atkinson showed Sculley the prototype, and the former sugar-water salesman was blown away. It was exactly the kind of world-changing innovation that he wanted to be associated with.

"What do you want?" he asked Atkinson.

"I want it to ship," said the Apple Fellow.

They cut a deal—Apple would either bundle HyperCard with every computer it sold, or grant full ownership of the product to Atkinson, so he could sell it elsewhere. Using his home as a development center, Atkinson led a team of several programmers in creating the product. It took almost two years, but by the time it was complete, HyperCard was a brilliant exploitation of the Macintosh's abilities. The program made it abundantly clear that the Mac was the first engine that could vivify the hypertext dream—one could hack out Vannevar Bush–like information trails by a few simple manipulations and clicks of the mouse.

John Sculley himself took great pleasure in introducing the product at the August 1987 Macworld Expo, calling it the most important product he had been associated with since coming to Apple. Later, he claimed design credit for many of its capabilities, including its ability to connect to optical media such as CD-ROM disks. HyperCard made quite a stir at the show. The development team, recognizable by their custom-made royal-blue bowling shirts, were accorded celebrity status. If you were lucky enough to run into Atkinson himself—finally accorded the Apple superstar status that he richly deserved—he would slip you a floppy disk with a special set of HyperCard stacks.

It seemed in some respects like the dawning of a new era, the age of hypermedia, where the common man would not only have tremendous access to previously elusive shards of information, but would actually become a master manipulator of that information. Danny Goodman, the computer journalist and technical writer whom Apple had chosen to write an official handbook to the program, was so inspired by this capability that he immediately wrote a complicated HyperCard stack to organize his life—address book, scheduler, etc.—with everything cross-linked. It was to be an example of this new era, when people with little experience or inclination to be computer programmers would actually be able to design sophisticated applications.

As it turns out, some of these dreams were overly optimistic. Even though the scripting language for HyperCard, called HyperTalk, was one of the least onerous high-level computer languages yet devised, it was still a programming language. For most people, programming a computer is fine if you can do it as painlessly as rearranging a desktop, inserting a formula in a spreadsheet, or redesigning a layout of a pinball machine by moving around the bumpers and slots—but not by learning a new language and syntax and actually writing programs. (The idea that programming a computer can be done metaphorically instead of in the traditional, painstaking manner is a major lesson taught by Macintosh.) So while thousands used the program to organize information, most of the millions of people owning HyperCard never bothered with HyperTalk.

On the other hand, HyperCard's mere presence on millions of Macintoshes was a subtle yet profound cultural milestone. Before HyperCard, those interested in realizing the dreams of Vannevar Bush and Ted Nelson had formed a community that, despite ardent efforts to educate people to their vision, remained on the fringe. These hypertext adherents had been regarded in the same way that linguists treat the fervent proponents of Esperanto—cultists who may have a point to make, but whose cause is doomed. The appearance of HyperCard, the product with hypertext as its heart, changed all that.

It was as if the Esperanto people had suddenly been presented with an entirely new culture who spoke Esperanto as their first language! "If you look back just two years ago . . . there were literally only two books available with any great mention of hypertext," said a speaker at the Hypertext '89 conference in Pittsburgh. (The books were Ted Nelson's.) "By 1989 there were a dozen books fundamentally about hypertext that I could gather from my office in under three seconds. And there are literally scores of others that purport to be about hypertext that are sprouting up like dandelions."

That was the intangible benefit of HyperCard—a hastening of what now seems an inevitable reordering of the way we consume information. On a more basic level, HyperCard found several niches, the most prevalent being an easy-to-use control panel, or "front end," for databases, providing easy access for files, pictures, notes, and video clips that otherwise would be elusive to those unschooled in the black arts of information retrieval. Thus it became associated with another use of Macintosh that would become central to the computer's role in nudging digital technology a little closer to the familiar: multimedia.

In recent years multimedia has taken on a negative connotation in the computer industry. The term is often used with a suspicious fuzziness, and is often dismissed as a meaningless buzzword, tainted by hucksters invoking the word to move new hardware. Stripped of its gloss, however, multimedia is simply this: the integration of the technologies of entertainment—stereo sound, animation, video—with heavy-duty information processing. Multimedia is the attempt to make computers behave like television sets, only interactively.

Macintosh, in its bit-mapped refusal to discriminate between text and graphic, was the natural platform for this nascent technology. It had the potential to bring the Macintosh magic to a much wider audience.

In a sense, mapping other media into computers was inevitable. Despite the advances of the Mac interface, a substantial percentage of the public, probably the majority, still mistrusted computers and had little desire to master them. Yet these same people spent a numbing proportion of their days transfixed by technology—the telephone, the stereo, the radio, the high-speed printing press, and above all, the television. Imperceptibly to the consumer, all of those "old" technologies were well on their way to transforming themselves. Their appearances seemed no different, but the signals that produced those appearances were changing to digital format, a lingua franca for experience in the next century. The day would come when, like snakes shedding skin, they would suddenly show their true colors and be transformed into a single multimedia exclamation point.

That day isn't here yet—but you can glimpse it by playing with Macintosh, circa the early 1990s. While you struggle to evoke prose from your word processor, punctuated perhaps by verbal notations injected by use of the built-in microphone that comes with almost every Macintosh, you can also devote part of your screen to viewing a complete movie—*A Hard Day's Night,* or a porno flick entitled *Further Adventures of Buttman.* And then you can capture a frame from the picture and place it in your document—or even append an entire scene to the document, storing it in a format developed by Apple called QuickTime. With the right software, you can even get a visual scan of the vocal message you just added and by fiddling with a few horizontal scroll bars you could change the pitch or even the content—maybe even changing the sound from a human voice to a Hammond organ. (The software could help you to digitally edit music produced on your desktop recording studio by use of a standard called MIDI, which was sort of a PostScript for synthesizers.) If you wish, it is even possible to tinker with the film's montage, rearranging the scenes, perhaps even intermingling them with video clips taken with your camcorder, reading them into Macintosh with an attachment called Video Spigot. You can place a photographic portrait on an Apple flatbed scanner, read the information into it, and using software with exotic names like Digital Darkroom, alter the reality of that photograph, either touching it up to remove a blemish or performing more drastic cosmetic surgery with a paint program. You can even dabble in what formerly was the sole domain of motion picture special effects—"morphing" one image to another. A software package called Morph selling for under $100 does this.

To be honest, very little of this is as easy to perform as the classic functions of Macintosh—it takes a little time and probably more than a modicum of brains to master the new skills. Yet by virtue of the Macintosh consistency, all these tasks are many times easier to pull off than they had been in their previous guises—expert systems costing thousands of dollars more. *The weird turn pro.*

What all these functions have in common is the manipulation of digital information to create a reality that seemed more a function of the natural world than of the alien computer realm. Macintosh's first step was to create the subtle virtual reality of its interface. Step two pulled the trappings of the outside reality into this virtual world.

Was there a logical destination that could be plotted by this trajectory? John Sculley seemed to think so. His beliefs and desires were reflected in a creation of his own. Like Kay's Dynabook (an obvious inspiration, especially considering that the Apple chairman regularly met with Kay), Sculley's creation was a fantasy construct that would never be tested by engineering or marketplace realities. It served two purposes—a glimpse into the future of Apple's products as sketched by the company's leader, and a bid for Sculley himself to be admitted to the pantheon of technological visionaries. The man who as a frustrated adolescent drafted doomed plans for a television tube would now have his visage alongside the other giants in the virtual Mount Rushmore of the information age.

He called his invention the Knowledge Navigator, "a future-generation Macintosh, which we should have early in the twenty-first century." As he wrote in his book, *Odyssey,* subtitled "a journey of adventure, ideas, and the future":

> Imagine the Knowledge Navigator having two navigational joysticks on each side, like pilot's controls, allowing you to steer through various windows and menus opening galleries, stacks, and more. You might even be set free from the keyboard, entering commands by speaking to the Navigator. What you see on the large, flat display screen will likely be in full color, high-definition, television-quality images, full pages of text, graphics, computer-generated animations. What you hear will incorporate high-fidelity sound, speech synthesis, and speech recognition. You will be able to work in several of these windows at any time, giving you the possibility to simultaneously compare, for example, the animated structural system of living cells with the animated network of a global economy. Or you might want to explore the depths of Zen philosophy in which beauty is in the details, comparing it with examples of the architectural details of the Parthenon from ancient Greece and then contrasting these ideas with the design details of a Japanese camera. Various windows on the display will give you a choice of text audio, animated graphics, or television-quality images, letting you simultaneously grasp ideas through a mix of media alternatives.

Sculley's Knowledge Navigator was a pastiche of Bush, Engelbart, Kay, Atkinson, and Dick Tracy. Every aspect of it had already been well traded in the marketplace of ideas; much of the enabling technology that would be required to realize the device was already being developed in laboratories, waiting only for Moore's Law to drive down the price. Yet Sculley won acclaim for his futuristic paste-up job.

This success was due in part to a series of long-form virtual commercials for the Knowledge Navigator. Talking to George Lucas one day, Sculley wondered aloud if it would be possible to exploit the kinds of cinematic special effects used in the *Star Wars* series to dramatize the Navigator. Thus was born a series of five-minute documentaries that simulated a product current technology could not realize. Apple's industrial design group mocked up a laptop-sized Navigator—it folded out like a bound notepad and had deco trimmings—as the repeating element in this anthology.

The first episode was typical. A Berkeley professor engaged his device in a colloquy intended to prepare his lecture notes for a class he would teach later that day. True to the politically correct Apple ethos, the class was to consider the phenomenon of global desertification triggered by destruction of the rain forest. The professor, a somewhat self-absorbed twit, really, addressed his comments to part of the display screen that depicted the image of a wry young man in a bow tie. He appeared to know everything about the life of the Navigator's owner and devoted his own virtual life to performing any digital chores the professor could conceivably require. If you wanted to be nice about it, you would call him a factotum. Less charitably, he was a slave made out of software. If for some reason he got out of line, his owner could drag him to the trash can and replace him with a more obsequious icon.

When the professor requested animated visuals that vividly illustrated the extent to which humanity was raping the rain forest, the Navigator quickly gathered the information from the hypertext libraries stored in humankind's vast digital archives, crunched the calculations, and extrapolated the results on a high-definition map that would shame *USA Today*'s graphics department.

But the strangest part of the video occurred when the bow-tied slave located a recent paper by Jill, an associate of the professor's. When he heard this, he asked his digital assistant to get her on the phone so he could ask if she'd speak to his class (remotely, of course). Anticipating his request, the Navigatoid had already contacted her. And poof! there she is on the screen, a real human being, appearing in a window underneath the square displaying the virtual twit.

At least we viewers *think* she's real. Though the video offers no direct evidence that forces us to consider this possibility, upon a second viewing I had the frightening realization that aside from some toothless bantering, Jill's appearance and behavior was not really more human than that of the bow-tied homunculus. What if she were not the professor's colleague, but Jill's own software agent? Macintosh teaches us to regard all sorts of media, from sound to text to video, as malleable building blocks, strands of thread we can weave into our virtual reality and change—will Knowledge Navigator teach us that *people* are but another medium?

I suspect that these musings are more the product of my own imagination than Sculley's vision. Such dark speculations were nonexistent in the chairman's speeches, where information technology appeared as a benign force solely devoted to making us consider things like the Zen-like principles of Greek facades and Nikon cameras. The future he presented was relentlessly bright.

"This is an adventure of passion and romance, not just progress and profit," said John Sculley at the 1988 Macworld Expo. "Together we set a course for the world which promises to elevate the self-esteem of the individual rather than a future of subservience to impersonal institutions. . . . the 21st Century Knowledge Navigator shows us [that] the journey to evolve the amazing Macintosh has only just begun."

Yet at the same time Sculley first began propounding his views on the Knowledge Navigator, Apple itself was failing miserably in bringing a portable Macintosh to market.

The first effort at this was called the Macintosh Portable. For over a year the company had whetted the appetites of Mac cultists, promising that Apple would deliver a "no compromise" package that contained all of the Macintosh magic in one easy-to-carry package. But by 1989, when Apple delivered the Portable, it had compromised the essence of the project—portability. In its attempt to deliver, undiluted, the experience of working with Macintosh on a desktop, Apple gave birth to a beige elephant. It weighed almost as much as the original Macintosh, and cost twice as much. It was the worst excess of the testosterone-charged Gassée era at Apple.

For a few weeks in the autumn of 1989 I lugged around a review copy with me when I visited the Bay Area. The Portable's main advantage was exercise; its density roughly approximated that of barbells. It was far too big to fit on an airline tray. I wound up trying to take notes on it while balancing it on my knees. In ten minutes, my circulation had stopped and in the interest of maintaining a good relationship to my lower body, I flicked off the Portable. Even when it was perched on a desktop, however, the Portable was an annoying computer. The designers had implemented all sorts of features to preserve the life of the lead-acid batteries that helped add to the machine's ponderous avoirdupois. Apparently no one had realized that at its present size, the Portable was so un-portable that it would rarely be transported to locations—the beach, Mount McKinley—where electrical outlets were not plentiful. (I did hear that an astronaut took it into orbit, a development that does little to affirm one's confidence in NASA.) Finally, the liquid crystal screen display, though fairly sharp for its time, did not have "backlighting"; it required outside illumination. The enormity of this lapse was apparent to me when, during the review period, I happened to be in the newsroom of the *San Francisco Examiner* on the evening of the big 1989 earthquake. The building's power was off, and the staff was hastily cobbling together the next day's edition by using portable computers. None were Macintosh Portables—

in the gloom, only machines with backlit screens could be put into service. (Ironically, the files were later translated to Macintosh format so that the earthquake edition of the paper could be desktop-published and printed elsewhere.)

It was not until 1991 that Apple was able to muster the wherewithal to compress the Macintosh experience into a notebook-sized computer. It was called the PowerBook, a conscious nod to Alan Kay's Dynabook. At one of the early Macintosh retreats, Steve Jobs had written on a blackboard, "Mac in a book in five years." He was only off by a few years.

In contrast to the Portable's out-of-touch design process, Apple directed the PowerBook's creators to observe the way people actually used technology. As a result, the designers felt flexible enough to toy with changing the shape and feel of Macintosh while remaining true to its spirit. The PowerBook has an entirely new look, different not only from both branches of its parent family—the compacts like the original Mac and the new Mac Classic, and the modulars like the Mac II and the Mac LC—but distinct from the laptops of its competitors. Most visually startling was the keyboard component. It occupied only the top portion of the base; below it was a roomy wrist-pad for ergonomic comfort and an easily handled track ball that performed the work of the mouse.

It was the best Macintosh since the original. Like its forerunner, it spun a compelling mystique. Coming to life with a distinctive chime, the clear, well-lit monochrome screen (backlit, of course) swept you into the familiar world of Macintosh. Yet something about the package—a stately gray exterior that cracked open like a clam, with keyboard on one side and screen on the other—made the experience even more intimate. Since it was sufficiently compact to carry on a crosstown jaunt, PowerBook people loaded it with very personal files—information relating not only to business matters but the detritus of their private lives. People wound up relating to it as if it were a diary. Inexpensive modems went inside PowerBooks to make them communications centers—digital telephones and fax machines. The cumulative effect was as though Apple had reached out into the void and bottled the future, as neat a trick as catching moonbeams.

Apple sold a billion dollars' worth of PowerBooks in its first year and could have sold more if its suppliers had been able to produce them. PowerBooks became status symbols. *USA Today* reported that Hollywood screenwriters literally slept with their Model 170's. *Newsweek,* in an article about the young Turks in the Clinton administration, reported that the battle cry in the White House was More PowerBooks!

One of the most interesting applications designed specifically for the PowerBook involved . . . reading. A company called Voyager published on floppy disks an ambitious list of books skewed to the Mac crowd (like William Gibson's cyberpunk thrillers), current best-sellers (the latest from Gloria Steinem to John le Carré) and classic works of literature (the works of Random House's Modern Library—Dostoyevsky to Eudora Welty). The "Expanded Books Project" flew in the face of continual scoffing from a literary establishment that had always regarded computers as the antithesis of art—but the original designers of the Macintosh would have found it a good match for their own work of art.

"This is . . . an effort to find a place for literature in the dynamic medium that's represented by the computer, rather than cede the future to MTV," Voyager's president Bob Stein said at a conference in 1993. "The goal at this stage [in the information age] was twofold. We wanted it to look enough like a book that readers would say, 'I recognize that. It's a book', and not throw up at the idea of reading on a screen. We also wanted to keep intact a book's functionality. We couldn't provide the fetishistic relationship to the object, and didn't want readers to lose too much, or they wouldn't like it."

As Stein admitted, the experience of reading Dickens or Ken Kesey on a PowerBook could not provide the tactile intimacy of ink on bound paper. But as with other successful software metaphors, these virtual books could outperform their traditional counterparts in other ways. Since the text of the book was stored as an electronic document, the reader could interact with it as if it were a word-processing document. Searching for words and phrases was instant, and any passage could be easily copied and dropped into an original document. Voyager also added new materials to each work, accessible, hypertext-style, by clicking at the proper link in the original document. During the reading of *Moby-Dick,* you could get a QuickTime video of a roiling ocean. Pictures of dinosaurs popped up in *Jurassic Park.*

It was yet another case in which Macintosh pointed

to—and hastened—a future where we might interactively consume and reshape all our information from a vast library of ones and zeros representing the accumulated wisdom, expression, and knowledge of humankind. Clearly, if Vannevar Bush had lived to be a hundred and two, he would have bought a PowerBook.

11

And now a confession. For much of the preparation of this book, ostensibly an homage to an insanely great instance of technology, I have been locked in battle with my protagonist. For weeks, something had been awry with my Macintosh IIcx (a 1989 successor to the Mac II). Every time I walked away from the computer for ten minutes or so, I would return to find the screen frozen. When I moved the mouse, the cursor icon—the little arrow that survived the journey from Xerox PARC to the Lisa to Macintosh—moved with it. But that was the sole response I could evoke. I could not type; I could not pull down a menu; I could not open an application by selecting an icon and double-clicking. The machine had "hung." The only thing I could do was switch off the computer and be thankful that I compulsively save my working files every time I wander away from my desk.

A properly working Macintosh is a marvel, but a Macintosh with an undiagnosed software problem is slow torture. Is the trouble caused by a faulty program?

There are more potential culprits on my desktop than suspects in a game of Clue. A virus? I use an "anti-viral" program to intercept these virtual parasites but even the best of these claim protection from only "all known viruses." Techno-anarchists in Bulgarian virus factories churn out new strains with distressing regularity, so the effort to stay current is as hopeless as a vow to avoid the flu. Then there are the sorts of system problems triggered by the sheer complexity caused by an ensemble of programs running simultaneously. A typical Macintosh has come to accommodate several applications running at once, in addition to as many as a couple dozen "system extensions." The latter are software programs that, upon start-up, lodge a tiny piece of themselves in the computer's memory in case they are needed—sort of camel with a nose under the tent. As with any family of independent-minded sorts attempting to occupy a limited space, conflicts can arise within a given ensemble of software files. Programmers try to avoid them, but they are a consequence of the sheer complexity that arises when more and more rowdy occupants are shoehorned into a single vehicle.

All of this comes as a consequence of Macintosh's success. The original Mac, while certainly not immune to system failures, was incapable of this sort of elusive malady. You barely had room for the operating system and a single application. The fail-safe method of dealing with a software problem was to make another copy of the disk that came with the application. If the problem persisted, you'd know it was faulty software and you would either call the developer to complain, and probably hear that others had already reported the bug, and a fix was on the way. If not, you would avoid that application. In any case, you would have identified the problem.

But then Macintosh grew. Hard disk drives increased the amount of storage from the equivalent of a large manuscript to something approximating the combined verbiage in a Borders bookstore. The Mac's memory chips were called upon to handle not just a single application and the system, but several applications, as well as the new class of "memory resident" system extensions. Meanwhile, the system itself had grown to Brobdingnagian proportions. In much the same way that a successful new business, as its income rises from thousands to millions to billions, finds itself mired in bureaucracy, the Macintosh software system, elegant interface and all, is somewhat overwhelmed by its new demands.

It's been said that the life span of a new personal computer is akin to that of a dog: each year of a computer's life is equivalent to seven human years. Using that analogy, a computer really doesn't mature until it is close to three years old, reaches its prime between the fourth and fifth years, and gets visibly long in the tooth after year eight. By the end of a decade, it requires artificial means of life support. Since the ecology of personal computers is rather harsh, however, very few computers are privileged to die natural deaths; the vast majority succumb to inborn disease or predators. In the entire (admittedly brief) history of the industry, the only machines surviving to old age have been the Apple II, the IBM PC and its clones, and Macintosh.

In order to keep Macintosh fresh and competitive,

Apple had to extend its abilities while not pricing itself out of the marketplace. Through the magic of Moore's Law, this was easily accomplished: the 1984 original Macintosh cost $2,500 and was all but unusable without a $400 external floppy drive. In exchange for that $2,900, Apple provided the buyer with 128K of internal memory and a total of eight-tenths of a megabyte of floppy disk storage. Nine years later, the lowest-powered Macintosh was called a Mac Classic II. It had an advanced processor that ran programs many times the speed of the original, came with 4 megabytes internal memory (thirty-two times as much as the original); a single floppy drive that not only held almost three times as much information per disk as the original but also had the ability to read disks formatted with information from DOS systems; and a built-in hard disk drive with 40 megabytes of permanent storage. It cost less than $1,000.

To accommodate that power, Apple had rewritten the system several times while attempting to maintain the same original intuitiveness of the original Mac interface. At a certain point, this became impossible. The paradigm of files within folders, for instance, was quite elegant and comprehensible when one had a few dozen documents—it was just like a file cabinet! But when users commonly had thousands of documents in their hard disk drives, things became immeasurably complicated, as if the imaginary file cabinet filled a gymnasium. After years of talking about it, Apple finally released a new version of the system and Finder, called System 7.0, which, among other things, dealt with this problem in part by introducing the concept of an "alias." This feature allowed you to create multiple icons for each document or folder. So if you created a letter to your boss about a tennis date, you could put an alias for the document in a Business Correspondence folder, a Tennis folder, and a Recent Work folder. The document's icon would appear in all of these, and double-clicking on it would open the file, regardless of whether or not the original actually "lived" in that particular folder. Once you learned how to use aliases, Macintosh life was a bit easier—but the point is that one *had* to learn how to use them. It was one of dozens of new twists that, in the aggregate, made Macintosh less intuitive.

Not to be bested by the system, the applications themselves had grown bigger and more Byzantine, in manners not anticipated when Macintosh was first conceptualized. A good example is Microsoft Word, which became the most popular Macintosh word processor after Apple stopped including free versions of MacWrite with every computer. Originally Word fit quite neatly on a single 400K floppy disk. It did its job with a minimum of distracting bells and whistles. But Word 5.0, released in 1992, was another matter. Microsoft had stuffed it with features, including a spell checker, a grammar checker, a thesaurus, a drawing program, footnotes, bullet charts, automatic indexing, page layout capabilities, a table-of-contents generator, drop caps, envelope addresser, a separate facility to generate mathematical notation, and a screen saver that filled the display monitor with spacey color graphics when the machine was not in use. Word 5 filled up five floppy disks, and installing it took up the better part of a morning. And Word was by no means the biggest program; I have seen some filling up *nine* floppy disks.

Macintosh had become so complicated that in 1991, not long after Apple shipped System 7, it released a new program called At Ease. It was essentially what is called a "shell"—an interface to the interface, used to protect novice users from getting frustrated by having to negotiate a complicated system best left to so-called power users. An Apple apparachik once explained it to me. "It's a friendly desktop that replaces the Finder," he said. "It's directed to the home and education market and for people new to the Mac. It helps ensure success for novice users."

I was deeply disturbed at this. After all, the Finder, the program that greets all Mac users, was the essence of the experience. It was what was supposed to proclaim through its intuitiveness and simplicity that Macintosh was different—this was a computer that was easy to use. "Are you saying," I asked the Apple guy, "that *the Finder isn't friendly?*"

He blanched. "I try to use the word 'friendlier.' "

There was no doubt that in ten years Macintosh had become much more powerful and utilitarian. But a price had been paid. How long would it be before it was time to put the dog to sleep?

In any case, this was the background to my Macintosh troubles: the computer had become more complicated than anyone had imagined. Obviously, one of those complications was making my life miserable.

The fail-safe solution would be to start from scratch —back up all my documents, then wipe the hard disk clean. Then, in an orgy of disk swapping not experienced since my original Macintosh, I could reinstall the system, reinstall all the applications, reinstall all the system extensions and utility program and games . . . and hope the same thing did not happen again.

But I could not face the prospect of this task. Not in the middle of, um, a book. Such an enterprise would take hours, maybe days. So I enacted a short-term fix. I began stripping the system of possible offenders, beginning with frills I hardly used, continuing with features I found somewhat useful but could live without, and ultimately, removing certain programs and utilities I found extremely helpful in my everyday work patterns, but knew that their presence could potentially cause the horrible situation I found myself in. I was stepping back in time, making the Mac emulate the simpler, though less useful, computer I had. As I wiped out Super Boomerang, Background Printing, On Location, and SpaceSaver, I pictured myself as Astronaut Dave in *2001,* determinedly yanking out the chips in the supercomputer H.A.L., with the uncomfortable feeling that I was deconstructing a personality. When I was finished, my Macintosh IIcx was not so atavistic as to sing "Daisy," but it was, in a Mac sense, no longer itself. On the other hand, it no longer hung.

I knew, however, I would have to eventually solve the problem for good, and some weeks later, I finally vowed to do so. The first step was some sleuthing. I reactivated the potentially offending pieces of software, one by one, waiting until the problem reappeared. Finally, I thought the culprit was identified: a program called On Location that lived under the pull-down Apple menu on the left side of the screen. Published by ON Technology, a company founded by Mitch Kapor after he left Lotus, this was a searching program that could speedily riffle through all the files on a hard disk to find any given phrase or word. I called ON's customer support line and reached a technician who informed me that On Location was normally very sound. But a recent version of a popular program had caused some problems.

"Do you have Microsoft Word version five-point-one?" he asked me. I sure did.

Put briefly, there was a kind of allergic reaction between On Location and the brand-new edition of Microsoft Word. Oddly, the source of the dispute could be mitigated if I accessed the folder that came with Word entitled "Sample Documents" located in a file called "Employment Report," and consigned it to the trash can. After this deletion, and removal of another possibly offending file he noted, things would quiet down, my support person predicted. I followed his directions and—poof, no more hanging.

In retrospect I marvel at how the problem arose and was resolved. It was as though having that particular file in my system, along with the preexisting program caused the same chaos as a butterfly flapping its wings in Indonesia. The machine's eventual breakdown was the hurricane in Kansas. Who knows what similar entanglements were emerging in the chaos of millions of lines of computer code silently executing on the motherboard of my IIcx?

Yes, the riddle was solved. But the experience jarred me into contemplation about my own relationship with Macintosh, and with technology itself. The first concerned the degree to which I had become dependent on it. What had become of the suspicious anti-warrior of the sixties, casting reproachful glances at the Temple University computer center? He had been turned. He was now a technophile. His compatriots from the rebel days now thought him a nerd. The first thing he did in the morning was switch on a Macintosh IIcx and the blackness of the screen would suddenly crystallize into the Finder, icons, windows, and menu bar on a sea-blue background. He swan-dived, cold, into this informational sea and spent the rest of the day bobbing in and out of it.

There was a time, I know, when I conducted much of the same sorts of business that I currently engage in, without requiring a machine that makes more calculations in a morning's work than all the combined arithmetical operations of humanity performed by hand, over the span of recorded history. But I can't remember what that time was like, or how I coped. It is only through the densest fog that I can even remember what it was like using my previous, pre-Macintosh, command-line-interface computer.

Still, I am hard pressed for proof that, for all its

magic, Macintosh has enabled me to be more productive. I *feel* that it has, with every inch of my being. But after my recent fiasco with On Location and Word, I sometimes question whether this is an illusion.

As it turns out, this question has been bedeviling economists as well. A few years ago Gary Loveman, a professor at MIT (who has since moved to Harvard), attempted to measure the productivity gains that came with the billions of dollars' worth of information technology purchased by American industry. Similar studies measuring the benefits of research and development had conclusively demonstrated that R&D was a solid investment, and that there was no reason to suspect that computer technology would be a different story. But when Loveman ran all the numbers, totaled the investments in information technology and then compared them to the productivity totals of the industries, he was startled, if not astonished by the results. "There was no positive effect," he said. "There may even have been a negative effect."

This gap between accepted reality (computers make us more productive) and the quantifiable result (they don't), has come to be known as the Productivity Paradox. A true puzzler: If computers enable us to get so much work done, in a much shorter period of time . . . why can't we measure it? Where did the productivity go?

Some people deny that the paradox even exists, and indeed, more recent studies indicate that the information technology investment *is* beginning to pay off—in part because of the ease-of-use of graphical interfaces like the Macintosh. Still, I think the paradox is a useful tool to assess the hours we spend focusing on our tools instead of using them—as in the better part of two days I spent trying to locate the source of my Macintosh troubles. This was a process in which I had never engaged back in the bad old days when I toiled on a typewriter. In a certain sense, those days were not bad at all. I never spent a whole morning installing a new ribbon. Nor did I subscribe to *RemingtonWorld* and *IBM Selectric User.* I did not attend the Smith-Corona Expo twice a year. I did not scan the stores for the proper cables to affix to my typewriter, or purchase books that instructed me how to get more use from my Liquid Paper.

But maybe productivity is not the main benefit from computers. As its designers understood, the point of Macintosh was not to prod you into piling up *x* more reams of paper, but to change the way you interact with information, to empower you to manipulate information with confidence, to augment your creative powers, and to change the very way you think. Macintosh has certainly expanded my view of information, enabling me to break the barrier between text and graphics. It has placed me into the slipstream of the digital age, probably even reorganizing my thought process to align itself to the point-and-click, cut-and-paste rhythms of Macintosh computing.

Some people have actually criticized Macintosh on the grounds that it does change one's thinking. Marcia Peoples Halio, an assistant professor of English at the

University of Delaware, raised considerable hackles among Macintosh adherents in the academic community by writing an article in *Academic Computing* entitled "Student writing: Can the machine maim the message?" In it, she described how the Freshman Comp papers she received differed according to the computers the students used to compose them. After several semesters of having her students use DOS-based IBM computers, she was shocked at the papers generated by Macintosh users: "Never before in twelve years of teaching had I seen such a sloppy bunch of papers," she wrote. The problem went deeper than punctuation: the Macintosh students wrote in a more casual style (Halio was reminded of the loose colloquialisms of the mass media) and even chose more frivolous subjects to write about. While the IBM students addressed issues like capital punishment and nuclear war, she complained, "Mac students chose to write about such topics as fast food, dating, bars, television, rock music, sports, relationships, and phenomena such as the foam 'popcorn' chips that come in many packages."

"Can a technology be too easy, too playful for young immature writers?" she asked. "It seems to me that schools with only Macintosh computers may need to alert teachers to the possible effects that using this icon-driven, super-friendly system can have on students' writing."

Twenty English professors at various institutions signed an indignant riposte directed to *Academic Computing*. The most pungent response, however, came from Stuart Moulthrop and Nancy Kaplan, associated with the writing programs at, respectively, Yale and Cornell. "To restrict argument to a narrow range of topics or to words alone may be defining 'composition' far too narrowly, especially in a world where information takes complex and sophisticated forms." They contended that Macintosh's features, especially the ability to integrate other media into documents, can "situate language—spoken, written, and iconographic—in a much richer context than the typed or word-processed essay can provide." In fact, they implied, the writing program that did not use Macintosh would doom its students to the backwater of an antique version of literacy. "Literacy in the next century may well mean the ability to compose in multiple discursive dimensions and across media . . ." they wrote. "If we English teachers are unwilling to expand our notions of writing, we relegate ourselves to the study of the past and the instruments of the past."

This was getting interesting—a showdown between the forces of tradition and the rebels with a graphical interface. But before the debate had a chance to develop, it became clear that it might become moot. Why? Because *all* computers were going to adopt the graphical interface. All of them were going to look like Macintosh.

During the 1980s, while some people were falling in love with Macintosh and others, including Marcia Peoples Halio and countless MIS managers in corporate America, were resisting its charms, the most powerful human being in the computer industry was working doggedly to ensure that all personal computers, regardless of whether they were Apple or its competitors, would work like Macintosh. This was, of course, Bill Gates, the chairman of Microsoft. Gates had been impressed from the start with Steve Jobs's original Mac presentation, and immediately launched a division of his company which would produce the most popular applications for Macintosh. But his larger goal was to replace his own lucrative product, DOS.

There were two reasons for this desire. The first was a genuine belief that the graphical interface was superior. The second was less altruistic. Microsoft had made its name, and much of its revenues, from operating systems and computer languages. But it wanted to dominate the applications market as well. It nettled Gates that Microsoft's competitors in the applications marketplace—the Lotus spreadsheet, the WordPerfect word processor, the dBase database—dominated those categories in the DOS domain. By ending the DOS era and wiping the slate clean, Microsoft could drive its own products to the top of the software best-seller charts.

Gate's first stab at a graphical interface came in 1983. The product was named Windows. It had more in common with the earlier PARC interface than with Macintosh—for instance, its windows were not overlapping, as were Lisa's and Mac's, but "tiled" on the screen, each one demanding a certain share of the display's real estate. Windows 1.0 was also incredibly slow, and hopelessly late. It did not catch on. Neither did Windows 2.3, the next major release. But in 1990, Microsoft released, with unprecedented fanfare, Windows 3.0, and found instant success. In the next three years, Microsoft sold twenty-five million copies of Windows 3.0, and by 1993 the Software Publisher's Association was reporting that sales of Windows software exceeded that of DOS and Macintosh combined.

In countless tiny ways and a number of significant ones, Windows 3.0 was inferior to Macintosh. It was hampered considerably by the fact that it was not designed from the ground up, but grafted on to the overburdened DOS system. (This was a necessity; otherwise it could not run on the millions of previously existing computers of that ilk.) All too often, the user could peer through the cracks and see the fault lines underneath. For instance, as with DOS, the names of Windows documents were limited to eight characters. And one could not bestow names upon individual disk drives and floppy disks. So while in Macintosh, one could name a file LETTER TO DEBORAH BRANSCUM 3/14/93 and save it to a disk named MARCH 1993 CORRESPONDENCE, with Windows the file would have to be named something like **BRNS_LET.DOC** and saved to **/B:**.

Yet for all practical purposes, Windows *was* Macintosh. Despite the lack of key components like the trash can, the essential operations—double-clicking icons to launch files, for instance—were identical. Applications written originally for Macintosh, like Excel and PageMaker, worked much the same on Windows.

For Apple, countering the perception that there was little difference between Macintosh and Windows was a difficult task. Throughout the 1980s it had commissioned a number of studies that proved Macintosh's superiority over MS-DOS, so now it attempted to do the same with Windows. It circulated a study by "a leading independent PC-testing company" called Ingram Laboratories that compared various models of Macintosh and corresponding IBM PCs or Compaq computers running the same software in Windows versions. The study claimed that the same applications performed better in the Macintosh models and that Macintosh offered better price performance. Leaving aside a shifty omission—the deck was stacked in Apple's favor by not considering the powerful PC "clones" among the competition—the very comparison showed that Apple had lost its chief arguing point. For years, it had been futile to equate Macintosh with its counterparts in the personal computer world—the comparison was truly, no pun intended, an apples-versus-oranges situation. But with Windows 3.0, the oranges had become apples.

In fact, by 1990—oh, irony of ironies—it was *Microsoft,* not Apple, which circulated results of a Temple, Barker & Sloane study (yet another "independent consulting and research firm") concluding that a graphical interface—in other words, Macintosh and its imitators—"generates a greater return on information technology investment than a traditional interface [like Microsoft's own DOS]."

In a very large sense, it was a great victory for Macintosh and the vision behind it. All computers were destined to be descendants to Mac. The lessons of Macintosh would percolate throughout the entire personal computing community.

Apple, however, would not be doing the teaching. The moment to capitalize on the Macintosh's superiority had been lost. In retrospect, it is clear that John Sculley should have opted for lower-cost Macintoshes several years earlier, and perhaps even widely licensed the Macintosh operating system to outsiders. Then Macintosh's market share would have been significantly more than its ten to fifteen percent of the personal computer industry. Instead of licensing the Macintosh, Sculley made alliances with other companies, agreeing to jointly develop future products.

The most startling of these was an agreement forged in 1991 to join with another large company to produce the next generation of computers. Apple's new partner was its former blood nemesis—IBM. IBM! To those of us who recalled the rhetoric surrounding Macintosh's introduction—Steve Jobs claiming outright that "IBM is out to crush Apple"—this shift was straight out of Orwell's *1984*, when the three global superpowers would shift alliances on a dime. Apple joining with IBM was like Luke Skywalker strolling off into the sunset with Darth Vader.

When Apple's profits dipped in 1993, Sculley wound up leaving his post as Apple's leader. The company's board of directors now considered him too much a visionary, an excessively starry-eyed technophile, to make the hard decisions necessary to shepherd Apple through the 1990s. Sculley was permitted to retain the title of chairman, but no longer had a direct role in the company's operations—exactly the fate of Steve Jobs in

1985. The irony was inescapable: originally hired to anchor Jobs's dreamy fantasies, John Sculley apparently departed because his technological seduction had become too complete—he was running Apple too much like Steve Jobs had and, like Jobs, he was gone within months—still attempting to make a dent in the universe but no longer at Apple. (In Sculley's case this new launching pad was a relatively obscure company involved in wireless communications.) Sculley's replacement was fifty-year-old Michael Spindler, whom Sculley himself had imported to Cupertino from his native Germany. A no-nonsense businessman who eschewed publicity and was not known to gush about 21st Century Renaissances, Spindler's nickname was "the Diesel." There was little danger that the Diesel would become Steve Jobs.

Apple had suffered down periods in the past, only to rebound. It was a company that sold $7 billion of worth of technology each year, and its product line still featured the best personal computer in the world, Macintosh. It would undoubtedly recover once again, perhaps with the fruits of its collaboration with IBM, a faster computer that Macintosh called the PowerPC. But another milestone had been passed. It seems that at Apple, eras, like computers, are measured in dog-years.

The changes at Apple bring me back to my dinner conversation with Steve Jobs in November 1983, just before Macintosh was introduced. He was talking about the future of Apple. "Something happens to companies when they get to be a few billion dollars," he said.

"They sort of turn into vanilla companies. They add a lot of layers of management. They get really into process rather than result, rather than products. They lose touch with their customers. Their soul goes away. And that's the biggest thing that John Sculley and myself will get measured on five years from now, six years . . . Were we able to grow to a ten-billion-dollar company that didn't lose its soul?"

"I think Macintosh accomplished everything we set out to do and more, even though it reaches most people these days in Windows," says Andy Hertzfeld today. "We loved the Apple II. And we loved art. So we made the Mac a descendant of the Apple II, and a computer for artists—for writers and musicians. We never doubted that the way we did things would catch on. The key thing is that we kept the Apple II spirit, the crazy irreverence, the anti-authority flavor. Macintosh tells people as they use it, 'You don't have to take things too seriously.' It was great to make a product that has a rebel heart."

He pauses. "And what I'm doing now is in that straight line, in that spirit."

What is Andy Hertzfeld doing now? Helping to create the spiritual successor to Macintosh. He is a cofounder of a start-up company called General Magic. Nestled in the cubicles of the General Magic engineering department is virtually a Mac Hall of Fame. In the corner sits Bill Atkinson, the corporation's chairman. Next to his cubicle is Hertzfeld, the inspiration to the young hackers on staff. Also in the room are Susan Kare, the original Macintosh artist; Dan Winkler, author of HyperTalk; and Bruce Leak, the key wizard behind Macintosh QuickTime. Joanna Hoffman, hired by Jef Raskin as a marketing person when only four people worked on Macintosh, is General Magic's vice president of sales and marketing. Even the communications director has a pedigree—she's John Sculley's former publicist.

General Magic began when Marc Porat, a former Stanford MBA working at Apple, came up with a product idea called Pocket Crystal—a personal communications device that would not only combine the functions of telephone, fax, electronic mail, but perform the tasks of a digital Filofax, and, finally, be a conduit to various databases and even, eventually, financial accounting systems. It would not only be an umbilical to the world's knowledge and a handy reminder of your personal business—it would be your wallet. The idea was so compelling that he was able to recruit Atkinson and Hertzfeld to join him. And the idea was so ambitious that the three cofounders convinced John Sculley to allow them to spin it off into a new company. Later, Porat gathered an alliance of partners that included Apple, AT&T, Sony, Motorola, Matsushita, and Philips.

The "products" of General Magic are not so much devices as technologies: Telescript, which is sort of a PostScript for communications—a computer language for so-called smart messaging; and Magic Cap, which is the pocket-sized, communications-based equivalent to the Macintosh interface. "The Macintosh," says Porat, is "the artist's sketch for Magic Cap." As with the Mac interface, the personalities of its key creators—Hertzfeld, Atkinson, and the art of Susan Kare—are readily apparent. The plan is to offer an interactive cyberspace of icons on a screen that fronts a hand-held device. For instance, the display might turn into a metaphoric music store. By touching the pictures of various shelves, one could browse through a stack of compact disks. Touching one CD icon might fill the screen with the label image. Touching again might trigger a wireless call to the record company—and the response would be a brief snippet of one of the songs on the CD. A button with a picture of a package might flash at that point, and if you touched it, the disk would arrive at your house the next day, billed to your bank account.

Oddly, among General Magic's companions in the pocket communications trade is a product produced solely by Apple itself. It's called Newton. Its head engineer is another Mac alumnus, Steve Capps. Working with him on the project are Macintosh classmates Jerome Coonen and Larry Kenyon.

Newton is a pen-based digital slab slightly larger than a human hand. Its icon-based interface is less whimsical than that of General Magic, but still an obvious successor to Macintosh. Example: when you scratch the pen over text on the display in an erasing motion, the text disappears—in a simulated puff of smoke. A metaphoric touch of Macintosh-style irreverence.

"With Macintosh, Apple was betting the company—

it was, 'If you don't pull it off, we die,' " Capps says. "With Newton, I always said 'Just make sure it's not a bet-your-company proposition.' Apple has its desktop business, so it's not like we won't be here [if Newton fails]. We have that freedom." Yet considering Apple's disarray at Newton's introduction in August 1993—Sculley no longer at the helm, stock price down, layoffs in the air—it seemed that Apple's future indeed rested with Newton and other instances of what it called "personal digital assistants."

These pocket communicators, or personal digital assistants, are not the only successors to Macintosh aimed at audiences that would ordinarily never venture near a computer. The next generation of television cable boxes as envisioned by Microsoft has a Windows-like point-and-click interface. Our remote controls will be pointing devices, allowing us to move the cursors over menus with entries like "Nightline" or *Casablanca.*

Further into the future, if we wear eye-phone goggles and other virtual reality apparel, the menus may appear before us in space, and the pointing device will be . . . our fingers. Sounds strange, but once I actually stood in a NASA laboratory and used my Data-gloved hand to invoke pull-down menus that shimmered before me like ghosts, and then chose the commands from these. It was as though I had stepped into the Macintosh metaphor.

Ultimately, we can expect to lose count of Macintosh's successors. Long after its departure, Macintosh will be remembered as the product that brought just plain people, uninterested in the particulars of technology, into the trenches of the information age. In the process it standardized the crucial bridge from metaphor to reality. The informationscape will never be the same. Neither will we. Armed with Macintosh understanding, with Macintosh knowledge, with Macintosh skills, we will cross the line between substance and cyberspace with increasing regularity, and think nothing of it. Many of us—millions in fact—already make the excursion daily.

All thanks to Macintosh, the computer with a rebel heart. Behold, a dent in the universe.

AFTERWORD:
The Making of the PowerMac

Ten years after my first exposure to Macintosh, I was back in Cupertino, meeting once again with a group of dedicated engineers trying to save Apple with a new computer. But it was not exactly déjà vu.

Apple's "campus" was now unrecognizable from the modest complex of the Jobs era. The faux adobe buildings on Bandley were abandoned; Apple now occupied a phalanx of fresh-off-the-assembly-line junior skyscrapers. Its latest headquarters, housing the executive offices and the Advanced Technology Group, was a complex just south of I-280. The building's address was One Infinite Loop. (This odd street name immortalized the oft-told joke about Apple's in-house Cray supercomputer, rumored to be so fast that it ran an infinite loop in six seconds.) Scattered on the lawn were huge sculptural representations of the icons one finds on the Macintosh desktop: cursor arrow, watch, and so on. A tangible tribute to the success of the Mac team's do-or-die gambit.

But as January 1994 approached, a cloud hung over Cupertino. Some said that Apple was doomed. While the ideas behind Macintosh had proliferated, the Mac itself had a measly twelve or thirteen percent of the marketplace. In order to survive, the pundits proclaimed, Apple would have to demonstrate clear superiority in the next generation of computers, run on more powerful microprocessors. And now was the time to do so. Once more, Apple had to put up or shut down.

Apple's response was a successor to the Macintosh. I was on a sentimental journey to see it and talk to its creators. My first glimpse of it provided nothing like the near-apocalyptic jolt I had experienced with the first Mac. The PowerMac looked almost exactly like the most recent Macintosh, the Quadra. And when Jim Gable, my guide to the PowerMac, booted the machine, I saw a desktop identical to the one on my home machine. It wasn't until he actually started up programs running in the PowerMac's "native" code—designed to run specifically on that machine—that the difference became apparent. Spreadsheet cells recalculated in a blur; 3-D images drew and redrew before the eye could register. The world had suddenly gotten faster. If Apple could exploit that speed, I thought, this could be the machine to carry it into the millennium. And save Apple in the process.

As with the Macintosh itself, Apple's new computer was developed in relative seclusion, at a conscious distance from the scrutiny of the company's leaders, who in any case spent the early nineties seemingly more interested in boardroom coups and distributing pink slips than in concocting nifty technology. Throughout the history of

Macintosh, it invariably has been the hands-on wizards who have always been able to come up with solutions before their leaders even understood what the problems are. And the PowerMac was no exception.

At the beginning there was a small group with a vision. It was led by Jack McHenry. In 1984, McHenry was a thirty-six-year-old Silicon Valley hardware gypsy. Then he saw the Macintosh, and, as he says, "from then on, my goal was to go to Apple and design Macintoshes." He got hired. After spending some time in Apple's disk drive division, he pitched Jean-Louis Gassée with an idea that was overdue: to make a Mac with a hard drive. He soon found himself at the reins of the Mac SE team. From there, he gathered a squad to produce the Mac IIfx, a 1989 success that was for a time Apple's mightiest machine.

Then, McHenry confronted the question that every manager of a computer design team faces when a project is completed: what next? The Macintosh IIfx, a costly beast nonetheless considered at the time a paragon of blazing speed and breathtaking power, was one of the first salvos in a fusillade that yielded about a zillion new Macintoshes over the next few years, an ascending spiral of more powerful models based first on the Motorola 68030 microprocessor, then the 68040. But some motherboard visionaries at Apple understood that that entire family of 68000 (or 68K) processors was essentially a dead end. In order to process cycle-hungry stuff like multimedia, telephony, and voice recognition—the stuff to provide the company's edge in the immediate future—Apple would need something even more powerful. A few people thought the answer was to switch microprocessors to something that used an exotic technology called *RISC.*

RISC is an acronym for reduced instruction set computer. Regular microprocessors, like those on the 68K family, have rich instruction sets that execute many commands. The inventors of RISC figured out that by cutting the instructions to a very few, you could have a faster processor. Sure, in order to perform the work of instructions that are no longer hardwired onto the chip, you have to run many more operations using the few instructions you have. But when the smoke clears, things still come out faster with RISC. Much faster.

Still, in 1990 RISC was a radical departure, a promising technology used so far only in very expensive workstations. Apple, as it turns out, already had a group working on RISC technology: the Jaguar group. The Jaguarians had decided that any Apple RISC computer would have to break cleanly from all previous Macintoshes. In their thinking, the first thing you would do upon switching from Mac to a new RISC machine was to feed all your software to a neighborhood goat—who would munch on it happily, since it was garbage. Because this machine would be powerful enough to do all sorts of heretofore impossible things, several futuristic technology groups at Apple allied themselves with Jaguar: projects like interfaces to telephones, audiovisual capabilities, and a voice recognition system. By 1990, there were between thirty and forty people working on Jaguar projects.

McHenry's team, however, approached RISC from a

different angle. On a group ski trip to Reno in March 1990, they made a decision to work on a RISC machine that would be a direct successor to the Macintosh family and run the current Mac software base. Considering McHenry's attachment to the Mac, this wasn't surprising.

McHenry's project was code-named Cognac, in honor of John Hennessy, a RISC pioneer with a surname identical to a particular after-dinner liqueur. The key people on his team included IIfx holdovers Bob Hollyer and Jonathan Fitch. The latter was also an old Apple hand who had even worked on the Lisa.

Cognac and Jaguar, of course, knew of each other's existence. They worked in different buildings, but each kept up on the progress of the other. And each thought the other team was hopelessly misguided. "We thought their approach was impossible," recalls Jon Fitch. Much of this parallel development took place during the Gulf War, and now, with some embarrassment, McHenry's team admits associating the Jaguar aggregation with Saddam Hussein. "We viewed Jaguar," says McHenry, "as the evil empire."

But the tension between the two RISC teams was small beer compared to the disquiet that these units sowed within Apple in general. It is easy to see why: There were literally thousands of people within Apple devoted to extending the current operating system—the world of the 68K—into the next decade and beyond. If either of these two small groups succeeded in moving Apple into the world of RISC, the company would see an upheaval not experienced since the Macintosh overthrew the Apple II. Though little-publicized, this drama was the real subplot of Apple's development efforts in the early 1990s.

The main problem with switching to RISC, of course, lay in the deep investment Macintosh users had with their software. One would expect the creators of high-volume applications like Microsoft Word or PageMaker to port their wares over to any new Apple platform. But a lot of Mac software falls into the realm of "folkware"—quirky applications that may not have won a large audience but help constitute the overall library that makes the Mac so valuable. At best, RISC versions of those applications would be slow in coming, and in many cases would never be attempted. If it couldn't run existing software, a RISC-based computer would be of limited value to Apple's current customers.

While running Macintosh software was not a priority for the Jaguar team, Cognac had to figure out how its computer could run both the 68K-based System 7 Mac software and the new generation of applications that would be created specifically for the RISC computer. This meant choosing between two alternatives.

The first was emulation, using the high-speed RISC chip to mimic the operations of the previously existing Macintosh. Emulation is no picnic. It often slows a machine down, and who wants to buy an expensive computer that makes your existing software look like you're working underwater?

The other option was to ship a two-in-one computer: a RISC machine with a 68K Macintosh chip set inside. But the history of such "dual processor" machines is littered with failures—no matter how elegantly you package them, they wind up as costly compromises. "Dual processors are a nightmare," says Jon Fitch. "So we were looking for a way not to include a 68K chip in the machine."

The breakthrough came late in 1990 when the Cognac team discovered what became known as the 90/10 rule. As McHenry explains, "It turned out that in a typical Mac application, ninety percent of the computing time is spent on ten percent of the code." This meant that it was theoretically possible to do a very fast emulator—it might actually be possible to produce a machine with only the RISC processor. "We could leave out the 68K!" says McHenry.

Meanwhile, the Jaguar team was shopping for a chip manufacturer to provide Apple's RISC processor. After a whirlwind tour of every potential RISC provider, the Jaguar team decided to use Motorola chips. (The rejected suitors were disappointed, since an Apple adoption of a company's RISC chip would mean a minimum ten-fold increase in sales.)

Both teams continued working independently. The main development in Jaguar was a change in code name: it became known as Tesseract. But Cognac's hardware design progressed nicely, and in late 1990, Cognac reached what's called "gray screen" (the point at which a prototype, its circuitry working, lights up a monitor). Still, the team knew that their efforts would be wasted if their Macintosh emulation was sluggish. "If there was a penalty [in the speed of preexisting applications] of buying this over the 68X Mac, we couldn't do it," says Jon Fitch.

The man charged with producing the Cognac emulation scheme was an engineer named Gary Davidian. Before coming to Apple, and working on various ROM toolboxes, he had been an expert microcoder—a person who works on microprocessor instruction sets. When Davidian joined the Cognac team, Apple had not chosen its RISC microprocessor yet, so every couple of weeks he was writing a new emulator, depending on the RISC chip *du jour*. It was not easy, making those chips pretend that they were 68Ks, but as he progressed he learned many tricks. By the time Apple settled on Motorola in 1991, Davidian was ready to use those tricks. By midyear, he had an emulator that used a RISC chip stuffed inside a Macintosh LC box. Look, Ma, no 68000! The RISCified LC, which they called RLC, ran off-the-shelf Mac software at speeds comparable to a Mac II.

"The RLC worked great—it blew away the company," says McHenry. It emulated almost everything that ran on the Mac. One particularly triumphant moment was the demonstration to the Tesseract team. The Tesseract engineers kept trying to come up with software that would break the emulator—they even dug up some ancient floppies, some from the days of the pre-Fat Macintosh, to stump the RISC machine. But RLC maintained the charade that it was a 68K computer. And when the RISC machine got to run software written specifically for its own machine—in what was known as "native code"—it churned out Mandelbrot fractals at dizzying speeds.

The existence of that prototype was a real milestone

ad to RISC. People could *see* how a dramatic p could actually open the door to a new style of puting—and at the same time they had living proof t the old software base would still work. "Until then, people didn't think it was possible," says Wayne Meretsky, who was then a Tesseract engineer. "They didn't understand that the limited instruction set could do all that."

But while the Cognac team produced more prototypes, Apple's executives were busy with another matter entirely. John Sculley and his colleagues had begun negotiations for a joint venture with, of all entities, its former blood enemy: International Business Machines. IBM. Darth Vader. Ironically, one of the things that had first brought Apple and IBM together was the Jaguar team's search for a RISC chip. Once that backchannel was opened, it turned out that Apple and IBM officers had plenty to talk about (undoubtedly a fear of Microsoft's dominating position in personal computers), and after months of top-secret discussions, in the summer of 1991, both companies made an announcement: they would work together on several projects. The most important of these would be the development, in conjunction with Motorola, of a new microprocessor that would be the heart of new machines from Apple and IBM. The chip would be called PowerPC. And it would be a RISC chip.

Apple people called it, with a measure of mockery, the Deal of the Century. As far as Cognac was concerned, it meant at the least a retooling of its emulation scheme—using a chip that literally hadn't been designed yet. Even more daunting, the team had to bring on software wizards who would develop an entirely new operating system built around this vapor processor. The PowerPC itself would be a collaboration between companies with cultures so different that some doubted that anything at all could come from it. (Picture Roy Cohn and Alger Hiss sharing a solder iron.) And all of this had to be completed to meet an incredibly tight deadline—January 24, 1994, the tenth anniversary of Macintosh!

According to the details of the agreement ironed out in November 1991, the first chips were scheduled to arrive in Cupertino in less than a year. Jack McHenry and his Cognac team were skeptical. All all, there had been all sorts of deadline problems with Motorola, Apple's chip provider in the past—and Motorola was just one company. How could a consortium attempting to combine such opposing cultures do better?

But in one of the first meetings between Apple and IBM engineers, something strange occurred. The IBM'ers dressed in blue jeans. The Apple-oids wore ties. Each had attempted to make the others feel at home. Obviously, everyone wanted badly to succeed here. And as it turned out, there was an open atmosphere at Somerset, the building in Austin, Texas, that became the design center for the PowerPC chip. "Once you get below the management level, highly talented engineers such as these just want to get the work done," says Paul Nixon, whom Apple recruited from Texas Instruments to head its Somerset team. Good thing, because there was a lot of work to do. IBM's 601 chip, the first in the PowerPC series, needed almost a "ground-up redesign," says Nixon. At the same time, engineers from three companies worked on the design for the next two chips, the more efficient 603 and the so-fast-that-no-adjectives-can-do-it-justice 604.

Back in Cupertino, the Cognac and Tesseract teams were working frantically to design the computers themselves. Since McHenry's team saw its machine as an affordable step between the current Macintoshes and the slick high-end Tesseract RISC machine—a "missing link"—the Cognac code name was changed to Piltdown Man (or PDM), after the famous, and fraudulent, attempt to identify the missing link in the evolution of humanity. While the higher-ups at Apple still saw Tesseract as the future, McHenry's team believed otherwise—Piltdown, by maintaining its Macintosh-ness, was the real road to future success.

The PowerPC 601 chip arrived in Cupertino on September 3, 1992, inside a package covered with Christmas wrapping. Somerset had made its deadline—almost unheard-of in the chip industry. The Piltdown prototype wasn't quite ready to go, but McHenry's team pounced on the chip anyway, immediately socketing it up to a circuit board and hooking it into a Mac. "On Monday I got the card," says Gary Davidian, the emulation expert. "Tuesday we powered up. On Wednesday we were trying to boot the Mac ROM. That night we broke something, so on Thursday we got a new card. Thursday night we booted Mac software. And on Friday we demo'd it."

On October 3, at approximately 5 a.m., the first prototype of a PowerPC Macintosh was booting the Finder. Unlike its namesake, this Piltdown Man was no hoax.

Things were not going as smoothly with the Tesseract. Put bluntly, the team couldn't get its ambitious machine, which was to be the high-end flagship of Apple's RISC effort, to work right. Around Christmas of 1992, the Tesseract team concluded that it could not possibly ship in the early 1994 time frame. Disaster. Even Apple's executives, who universally had favored the machine over Piltdown, had to consider whether the project might be beyond redemption. And in March 1993 they finally pulled the plug on Tesseract. Apple was only a year away from shipping PowerMacs, and, as Jack McHenry notes, "there was no high-end product." The whole PowerMac effort hinged on getting a replacement—fast. Apple's only hope, really, was if Piltdown Man could somehow be mutated into several versions, including a more powerful, feature-laden model.

"We had thought about that," says Jack McHenry, revealing a coy political acumen that comes from ten years of computer design at Apple. All along, it seems, McHenry's squad had designed its Macintosh so that the processor could be expanded—into the very sort of high-end machine that their rivals had been working on! So when Tesseract dropped the ball, there was McHenry and company, perfectly positioned to snatch it out of the air.

With a new deadline of March 1994, McHenry's team began transforming the humble Piltdown into three models. The first was the original PDM. The high-end version, maintaining the theme of scientific fraud, was dubbed Cold Fusion. Then a mid-range version was added, and code-named Carl Sagan, perhaps in honor of the billions and billions of dollars Apple might reap from it. (The eminent astronomer felt that this tribute somehow exploited him, and he sicced his lawyers on Apple. McHenry and crew then changed the code name—to BHA. Was it coincidence that the letters formed an abbreviation for Butt Head Astronomer? Sagan didn't think so. He sued Apple.)

Once the hardware issue had been settled, the PowerMac focus turned to its software. In effect, Apple had to fast-forward into the next century with a successor to System 7. To manage the effort, Apple tapped Phillip Koch, a RISC expert who had been doing research at Dartmouth. But Koch had difficulty recruiting Apple's top engineers for his team. "It was hard to get excited about building system software for a chip that was going to be in an IBM computer as well as Apple's," he says. "And it was also a case of 68K chauvinism."

Jim Gable, who had moved over from the Tesseract team to manage the entire PowerMac effort, was worried. "The software estimations were overly optimistic. We had a prototype, but very little native code running." His fears were confirmed when Sheila Brady, a brilliant and profane engineer, was recalled from a sabbatical at Harvard to beef up the PowerMac software effort. She took one look at the massive chart with various projected delivery dates and judged the deadlines as about as realistic as cold fusion itself. And Brady knew something about missing deadlines—she had been in charge of the System 7 development team that had set new standards in vaporware. "It was time to get *rigorous*," she says.

Koch and Brady methodically gathered a team and began to resolve the key issues in the PowerMac operating system. The most important question was whether it would use the powers of the machine to integrate new features. Would it *look* different than System 7? "The more we thought about it, the more we decided that the transition to a new interface would be too abrupt," recalls Jon Fitch. "After all, we're building *Macintoshes*."

Adding even one whizzy new feature to the PowerMac interface was judged too drastic. "So we ended up burying the differences between the Mac and the PowerMac," says Brady. Underneath the surface, the new operating system was drastically rewritten. Yet on the screen the interface was identical to System 7—so much so that the designers had to hack an Init program that helped them determine at a glance whether their prototypes were running the PowerMac operating system or System 7 (if a certain pixel was lit, it was the new OS).

The software turned out to be a dual success. True to the intentions of the Cognac team, it ran the current Mac software at speeds comparable to Quadras. Meanwhile, its familiar trappings cleverly shadowed the fact that this operating system would one day accommodate the innovations first conceived of by the now-defunct Tesseract group—telephony, voice recognition, multimedia, and all sorts of exotic interface improvements.

"We began with a core of five to ten software people and turned the company around," says Phil Koch. "It's clearly obvious that this is going to be the future at Apple."

Though the team missed the original deadline for the PowerMac—January 1994, a neat decade since Steve Jobs appeared on the stage at De Anza College with an odd-shaped canvas bag—they didn't miss by much. On March 14, 1994, at Lincoln Center in New York City, the PowerMacs had their debut. Apple had pretty much preannounced all the details of the three new computers it was introducing, so the event itself was something of a formality. Instead of the exhilaration of pushing the envelope of technology, the Apple executives seemed relieved that they had once again dodged the bullets of oblivion, by releasing a product that kept the company in front of the competition.

The legacy of Macintosh had been extended, but Apple's future was still dependent on a continuing streak of advances. Could it continue? No one knew. But for the weary Real Artists who made the PowerMac, the spirit of Macintosh was still a living force.

"I came to Apple to make Macintoshes," says Jack McHenry, bridling at the implication that the company had lost its soul. "And on its worst day, working for Apple is better than working anywhere else." The success of the PowerMac will gice him the chance to make another version of the machine that dented the universe.

BIBLIOGRAPHY

The main source material for *Insanely Great* has been my ten years of reporting on the Macintosh, mainly for monthly columns in *Popular Computing* (1983–85) and *Macworld* (1986–present). I have also reported on Macintosh and related technology for *Harper's, Rolling Stone,* the *Whole Earth Software Review,* the *Whole Earth Review,* and *Wired.* During that time I have had conversations and interviews with almost every major figure in the Macintosh community. In addition, I did a series of "debriefing" interviews with some key sources in the spring of 1993. Publications I've consulted include the journals devoted to the Macintosh: *MACazine, Macworld, MacUser, MacWEEK, Macintosh Today,* and *St. Mac;* other computer publications: *Academic Computing, Communications of the ACM, Byte, Hyper Pub, InfoWorld, PC Jr, Popular Computing, RELease 1.0;* general magazines: *Newsweek, Newsweek Access* (interview with Steve Jobs), *Playboy* (interviews with Jobs and John Sculley), *Scientific American,* and *Time;* and newspapers: the *New*

York Times, Wall Street Journal, San Jose Mercury. Below are books that provided information and ideas, any of which would be useful for further reading.

Apple Computer, Inc. *Human Interface Guidelines: The Apple Desktop Interface.* Reading, Mass.: Addison-Wesley, 1967.

Benedict, Michael, ed., *Cyberspace: First Steps.* Cambridge, Mass.: MIT Press, 1991.

Brand, Stewart. *The Media Lab: Inventing the Future at MIT.* New York: Viking, 1987.

———. *Two Cybernetic Frontiers.* New York and Berkeley: Random House/Bookworks, 1974.

Clapp, Doug. *Macintosh! Complete.* North Hollywood, Calif.: Softalk Books, 1984.

———, ed. *The Macintosh Reader.* New York: Random House Electronic Publishing, 1992.

Gassée, Jean-Louis. *The Third Apple.* New York: Harcourt Brace Jovanovich, 1987.

Gibson, William. *Neuromancer.* New York: Ace Books, 1984.

Heckel, Paul. *The Elements of Friendly Software Design.* Alameda, Calif.: Sybex, 1991.

Kawasaki, Guy. *The Macintosh Way: The Art of Guerrilla Management.* Glenview, Ill.: Scott, Foresman and Company, 1989.

———. *Selling the Dream.* New York: HarperCollins, 1992.

Lammers, Susan. *Programmers at Work: Interviews.* Seattle: Microsoft Press, 1986.

Laurel, Brenda, ed. *The Art of Human Interface Design.* Reading, Mass.: Addison-Wesley, 1991.

Levy, Steven. *Hackers: Heroes of the Computer Revolution.* New York: Doubleday/Anchor, 1984.

Manes, Stephen, and Paul Andrews. *Gates.* New York: Doubleday, 1993.

Moritz, Michael. *The Little Kingdom: The Private Story of Apple Computer.* New York: Morrow, 1984.

Nelson, Ted. *Computer Lib/Dream Machines.* Rev. ed. Seattle: Microsoft Press, 1987.

———. *The Home Computer Revolution.* Self-published, 1977.

———. *Literary Machines.* Self-published, 1981.

Norman, Donald A. *The Psychology of Everyday Things.* New York: Basic Books, 1988.

Nyce, James M., and Paul Kahn. *From Memex to Hypertext: Vannevar Bush and the Mind's Machine.* San Diego: Academic Press, 1991.

Rheingold, Howard. *Tools for Thought.* New York: Simon & Schuster, 1985.

———. *Virtual Reality.* New York: Simon & Schuster, 1991.

Rose, Frank. *West of Eden.* New York: Viking, 1989.

Sandberg-Diment, Erik. *They All Laughed When I Sat Down at the Computer.* New York: Simon & Schuster, 1985.

Schmucker, Kurt J. *The Complete Book of Lisa.* New York: Harper & Row, 1984.

Sculley, John. *Odyssey: Pepsi to Apple . . . A Journey of Adventure, Ideas, and the Future.* New York: Harper & Row, 1987.

Smith, Douglas K., and Robert C. Alexander. *Fumbling the Future: How Xerox Invented, Then Ignored, the First Personal Computer.* New York: Morrow, 1988.

Tognazinni, Bruce. *Tog on Interface.* Reading, Mass.: Addison-Wesley, 1992.

Young, Jeffrey S. *Steve Jobs: The Journey Is the Reward.* Glenview, Ill.: Scott, Foresman and Company, 1988.

INDEX